RAF DUXFC

This book is dedicated to the thousands of British, Commonwealth, Czechoslovakian and American personnel who served at Duxford during two world wars and those who sacrificed their lives to keep the world free.

Also dedicated to my mother Muriel, my wife Kim and mother-in-law Doris

RAF DUXFORD
A HISTORY IN PHOTOGRAPHS
FROM 1917 TO THE PRESENT DAY

Richard C. Smith

GRUB STREET • LONDON

First Published in hardback in 2006
This Edition first published in 2009
Reprinted 2016, 2017

Grub Street Publishing
4 Rainham Close
London
SW11 6SS

British Library Cataloguing in Publication Data
Smith, Richard
    RAF Duxford : a history in photographs from 1917 to the present day
    1.Duxford Airfield (Great Britain) – History – Pictorial works
    I.Title
    358.4'17'0942657

ISBN 978 1906 502 331

Typeset and designed by Roy Platten, Eclipse – roy.eclipse@btopenworld.com
Printed and bound in Great Britain by 4edge Limited, Essex

CONTENTS

ACKNOWLEDGEMENTS

The author would like to thank the following men, women, museums and institutions listed below, for without their help this book would not have been possible.

Mr Ashley Arbon MBE
Mr Peter Arnold
Mr John Beynon
The late Air Chief Marshal Sir Harry Broadhurst GCB, KBE, DSO, DFC
Squadron Leader Peter Brown AFC
Mr Gerry Bye, University of Cambridge Library
Mr John Cook (USA)
Mrs Diane Crain
Ron Dupas
Mr Jim Garlinge, Old Duxford Association
Mr Alec Gray
Mr Colin Griffiths
The late Flight Lieutenant Leslie Harvey
Squadron Leader Iain Hutchinson TD
Flight Lieutenant Richard Jones
Mr Edward McManus, Battle of Britain Historical Society
Mr John Munro
Mr James Munro
Ms Yvonne Oliver IWM
Group Captain Herbert Pinfold
Mr Peter Pool
Mr Peter Randall, 8th American Air Force Little Friends website

The late Robert Rudhall
Mr Andy Saunders
Mr Stephen Saunders (ASA Productions [UK] Ltd)
Mr Martin Sheldrick
Mr Peter Silver
Colin & Rose Smith (Vector Fine Art)
Mr Tom Stevens (ex 19 Squadron)
Mrs Edith Thomas
Jacques Trempe
Johan Visschedijek
The Photographic Archive of Imperial War Museum, London
The University of Cambridge Library, Image Service
The National Archive, Kew
The Old Duxford Association
The Battle of Britain Fighter Association
The Battle of Britain Historical Society
The Air Historical Branch, Bentley Priory

Thanks also to all my family, friends and aviation associates. To all those members of the public who continue to support my research, books, DVDs etc. To Vector Fine Art for their outstanding book launches and hospitality.

And finally to John Davies, Anne Dolamore and the staff at Grub Street for their continued support of my work.

SELECT BIBLIOGRAPHY

The following books and documents are of interest to those wishing to know more about the history of RAF Duxford.

National Archive Documents
RAF Duxford – Air 28/232 1924–1943
             – Air 28/233 1945
             – Air 28/234 Appendice 1940–1943
             – Air 28/1017 1946–1950
             – Air 28/1186 1951–1955
             – Air 28/1493 1956–1961

No. 19 Squadron Operations Book – Air 27/252
No. 242 Squadron Operations Book – Air 27/1471
No. 310 Squadron Operations Book – Air 27/1680

Battle of Britain Then & Now, Winston Ramsey, After the Battle, 1980
One of the Few, G/Capt John. A. Kent DFC, AFC, William Kimber & Co Ltd 1971
Flying Colours, The Douglas Bader Story, G/Capt P. B. Lucas, CBE, DSO, DFC,
    Hutchinson & Co Ltd, 1981
Honour Restored, S/Ldr Peter Brown AFC, Spellmount Publishing Ltd, 2005
Fighter Squadrons of the RAF, John Rawlings, MacDonald & Co, 1969
Airfield Focus Volume 1, Andrew Height, GMS Enterprises, 1982
Duxford Diary, USAAF, W.Heffer & Sons Limited, 1945

INTRODUCTION

Today the name of Duxford is immediately recognisable to many as the home of the largest collection of war and peacetime vintage aircraft on show to the public in Europe. It is also renowned for its annual air shows held throughout the year, which bring enthusiasts from all over the world to marvel at the flying aircraft dating from over seventy years ago.

Most people already know of Duxford's role during the Battle of Britain, when it served as the main aerodrome in 12 Group. However, this pictorial book tells of the aerodrome's ninety-year history dating back to the First World War and also of its sister aerodrome at Fowlmere, just a few miles away which was used as a satellite during WW2.

Duxford was an important base through the 1920s, when the Royal Air Force began to expand and the 1930s when the aerodrome was at the forefront of aviation technology, with aircraft changing from biplanes to the new RAF monoplane eight-gun fighters. Indeed, squadrons based at Duxford received the first Spitfires to go into service with the RAF. Its pivotal role during Dunkirk and the Battle of Britain is also covered as is the period during 1941–42, when it was base to the Air Fighting Development Establishment and Air Gun Mounting Establishment who were testing many new aircraft and weapons.

Between 1943–1945, Duxford became home to the American 8th Army Air Force's 78th Fighter Group, who flew missions escorting and protecting the massive daylight raids against the German industrial installations of the Ruhr.

After the Second World War had ended the Royal Air Force returned to reclaim the aerodrome and it was used as a jet fighter base throughout the late 1940s and through the 1950s until its final closure in 1961, when the RAF finally moved out.

Saved from destruction during the late 1960s, the aerodrome was used as a storage facility by the Imperial War Museum, who laid plans for expansion of the site. In 1967 it was used famously for filming the war movie epic 'The Battle of Britain'. Since first opening again to the public in 1976, thousands upon thousands have visited Duxford and continue to do so.

The photographs in this tome have been drawn from various sources, some official, others from private albums, taken by pilots, ground crews and WAAFs. All are equally important images showing the Duxford of yesteryear and today. May its future be secure.

Richard C. Smith
June 2006

COMMANDERS THROUGH THE YEARS

No. 2 Flying Training School

Wing Commander Wilfred K. Freeman DSO, MC	June 1920
Wing Commander Sidney Smith DSO, AFC	1st November 1921

RAF Duxford

Wing Commander, the Honourable L.J.E. Twisleton–Wykeham–Fiennes	10th July 1924
Group Captain C.R.S. Bradley OBE	8th August 1924
Wing Commander P. Babington MC, AFC	20th July 1925
Wing Commander R.G.D. Small	23rd September 1925
Wing Commander R.J.F. Barton OBE	11th February 1927
Squadron Leader H.W.G. Jones MC	12th May 1928
Squadron Leader E.C. Emmett MC, DFC	11th August 1928
Wing Commander F.L. Robinson DSO, MC, DFC	11th February 1929
Wing Commander N.C. Spratt OBE	29th September 1930
Wing Commander W.H. de Waller AFC	11th June 1933
Wing Commander E. Brownsdon Rice	30th December 1935
Wing Commander H.P.L. Lester	3rd August 1938
Wing Commander A.B. Woodhall OBE	21st March 1940
Group Captain G.H. Vasse	28th March 1941
Group Captain A.W.B. McDonald AFC	1st July 1941
Group Captain J. Grandy	12th February 1942
Wing Commander D.O. Finlay	19th December 1942

78th Fighter Group – 8th American Army Air Force

Colonel A Peterson	1st April 1943
Lieutenant Colonel M.F. McNickle	3rd July 1943
Colonel J.J. Stone	31st July 1943
Colonel F.C. Gray	22nd May 1944
Lieutenant O.E. Gilbert	1st February 1945
Colonel J.D. Landers	20th February 1945
Lieutenant Colonel R.B. Caviness	1st July 1945

Duxford handed back to RAF Fighter Command on 26th November 1945

Wing Commander A.C. Deere	1st December 1945
Wing Commander H.P. Burwood	9th August 1946
Wing Commander H.M. Pinfold	4th May 1948
Group Captain R.N. Bateson DSO, DFC	17th May 1951
Group Captain J. Rankin DSO, DFC	12th January 1953
Group Captain D.F. Macdonald	15th December 1954
Group Captain H.M. Pinfold	25th April 1956
Group Captain E.N. Ryder DFC	11th August 1958
Group Captain A.L. Winskill OBE, DFC	16th September 1960
Wing Commander J.H.S. Broughton DFC, AFC	1st August 1961
Squadron Leader P.E.A. Carr	1st October 1961

CHAPTER ONE
FLEDGLING EAGLES
1917 – 1929

After the outbreak of the 20th century's first global and most destructive war on 8th August 1914, a new aviation technology, the aeroplane was now being turned to use by the opposing Allied and German forces in the race for superiority. Initially used as a reconnaissance tool for the ground forces to supply vital intelligence, the early aviators began to arm themselves with pistols, rifles etc and began to take pot shots at one another in the skies over France and Belgium. It would be taken a step further when the aircraft incorporated machine guns, thus making the flimsy canvas biplanes into fighting machines that would engage one another in aerial combat.

The Royal Flying Corps was formed and many airfields began to be developed in Great Britain in preparation for either Home Defence or as training airfields for the new enlistment of recruits who wished to join the Royal Flying Corps.

A plan to build new airfields was quickly put into operation in 1916, in response to the new German threat from the skies of the gigantic Zeppelin airships, who now seemly unopposed began bombing towns and cities from the

A picturesque photograph of Duxford village green. This image was taken in 1913, just a year before the outbreak of the First World War. *(Author)*

Midlands down to London and the south-east. New aerodromes were now needed to fulfil the urgent requirement to train more pilots.

One site, in Cambridgeshire, which was surveyed for possible use as a training station by the Royal Flying Corps, was just north-west of the small village of Duxford. A survey team was sent to the area by the War Ministry and their findings reported a potential area of 223 acres on the south side of the Newmarket to Royston Road with an additional fifteen acres on the north side. This land belonged to Temple Farm and the owners of College Farm and after agreement it was then requisitioned by the War Ministry. Tenders were invited from construction companies to build the airfield and the contract was finally won by P & W Anderson Limited, a Scottish civil engineering company from Aberdeen during the summer of 1917, with work beginning on the site that October. The estimated cost of building the airfield was to be in the region of £90,000. The plans laid out for the aerodrome consisted of building three double Belfast hangars and one single with concertina shutter doors at each end, and various buildings for accommodation and stores.

A Glorious "Fourth."

Chronicle Photos.]
Independence Day was celebrated in Cambridge by a large number of American troops, who assembled in the town early in the morning and spent a right royal day. (1) The baseball match on Fenner's. The striker has just hit the ball (which can be seen travelling at a great pace out of the diamond). (2) The winning baseball team from Duxford. (3) The American Flag flying over the Great Eastern Station, Cambridge, as a welcome to all Americans coming into the town. (4) After the game at Fenner's, Col. Edwards calls for cheers for the Union Jack and Stars and Stripes, which were given in real American fashion.

American servicemen of the United States Aero Squadrons who were based at Duxford, celebrate 'Independence Day' on 4th July 1918 in Cambridge, where they held a baseball match at Fenner's. *(University of Cambridge)*

It is of particular interest to note that a second airfield was also being built just three miles down the road north from Duxford. Fowlmere had begun life as a landing ground in the autumn of 1916 for the use of 75 (Home Defence) Squadron against possible Zeppelin raids. It was also now to undergo construction with permanent hangars and buildings. The airfield was given one double and a single Belfast Truss hangar to accommodate aircraft. Fowlmere received its first RAF units on 1st March 1918, when 124, 125 and 126 bomber squadrons arrived from Old Sarum with DH4, DH6 and DH9 biplanes.

The work at Duxford however, with construction of brick and wooden buildings, progressed slowly during the next following months. It must be remembered how rural the area was at this time. Material was transported by rail to the nearby station at Whittlesford and then loaded on to horse-drawn wagons or lorries to make their slow procession to the site. By early 1918, it was estimated that the work force had grown to one thousand and that the price of the project had now gone well over the estimated cost. Nothing changes.

During March 1918, the site saw the arrival of 200 American personnel of 137th and 159th United States Aero Squadrons, who were to use the airfield as a temporary base for training. They were assigned to erect Bessoneaux hangars which would house the biplane aircraft temporarily while the permanent hangars were being constructed. Soon after the temporary hangars had been erected the airfield was put into use as a mobilisation station in March with the arrival of 119, 123 and 129 Royal Flying Corps Squadrons flying DH9s. (The service name of the Royal Flying Corps was officially changed on 1st April 1918, when it became known as the Royal Air Force.)

On 4th July, American servicemen celebrated Independence Day at the airfield particularly by staging a special baseball match at Fenner's in Cambridge which drew many local residents to the event, who waved both Union Jack and Stars and Stripe flags with great enthusiasm.

By mid 1918 more American airmen had arrived with 151st, 256th and 268th Aero Squadrons, but their role was limited to mainly assembling aircraft or running the motor transport section.

The three Royal Air Force squadrons were disbanded between July and August that year and the airfield's new role was as No. 35 Training Depot Station for Royal Air Force pilot training, using aircraft such as De Havilland DH4s, DH6s, DH9s, BE2cs and Avro 504Ks. Fowlmere was designated as No. 31 TDS, both aerodromes being under the control of 26 Wing at Cambridge.

Although work on finishing the airfield was still incomplete, Duxford was officially opened in September 1918 with the final cost an incredible £460,000. So outraged were the War Ministry that an inquiry was undertaken to find out where the money had been squandered.

By war's end on 11th November 1918, the training depot complement of staff had risen to 450 men with 158 women. The appointment of women in the newly formed Women's Royal Air Force, was an important step for the fairer sex. They undertook many roles that previously had been done by their male counterparts. Although in the main this comprised clerical duties, many women became mechanics, drivers or motor despatch riders, and some helped with the doping and stitching of fabric for aircraft.

The American contingent of airmen, no

longer needed now that peace had come, soon began to pack their belongings and head to the ports to board ships for the long trip back to the United States.

During this period many of the airfields throughout the country began the process of closing down with the land being given back to their previous owners. Fortunately for Duxford, it was decided that the airfield should remain as a permanent Royal Air Force station and be continued for use for pilot flying training. As a result of deciding to keep Duxford, a further survey was carried out in order to update the existing site and this resulted in extra work on re-roofing the hangars, construction of further buildings needed for expansion, as well as refurbishment of those already in use. The cost of this venture would be £355,000, or so it was estimated.

In July 1919, the airfield saw the arrival from France of 8 (Bomber) Squadron flying Bristol F2bs for disbandment, which finished in January 1920.

The airfield at Fowlmere, however, was put on care and maintenance for disposal at the end of March that year, being used as a storage depot for Handley Page bombers up until 1922, when it was finally closed and handed back to the landowner. In October 1922 demolition of the site was started and undertaken by the company, Bennett and Blowers, which raised the site to the ground, leaving little trace of the former airfield.

In April 1920, Duxford was designated as No.2 Flying Training School whose commanding officer was Wing Commander Wilfred Freeman DSO, MC. By June that year the unit consisted of twenty-seven officers, 390 other ranks and twelve pupils. The complement of aircraft was eighteen Avro (mono 504 Ks),

eighteen Bristol fighters and four DH9As. Wilfred Freeman stayed with the unit for just over a year before his position was taken by Wing Commander Sidney Smith DSO, AFC on 1st November 1921.

The training was varied and, apart from flying, included lessons in aerial photography, where either hand-held or fixed cameras bolted on to the side of the aircraft were used. A team from Duxford was entered into the photographic competition for the Royal Air Force Pageant held at Hendon in 1921 and achieved top place, with second place going to the School of Photography, Farnborough.

The training of aircrew continued, but the powers that be now decided that Duxford's role would also be that of a fighter station. This was realised on 1st April 1923, when 19 and 29 Squadrons reformed and were given Sopwith Snipe aircraft. Later that October another squadron, 111, arrived to complement the two other fighter squadrons and exchange their Snipe aircraft for Gloster Grebes.

The RAF then decided to move No. 2 Flying Training School to another site and they left for Digby, in Lincolnshire in June 1924, after four years of invaluable service with a grand total of 8,385.50 flying training hours achieved. On 23rd June, the aerodrome witnessed the formation of the station headquarters at RAF Duxford.

All three squadrons at Duxford had converted to the new Armstrong-Whitworth Siskin biplanes and this year also saw the first official station commander, Wing Commander L.J.E. Twisleton-Wykeham-Fiennes who took up the position on 10th July. It was an extremely short command, as by 8th August he had been replaced by Group Captain C.R.S. Bradley OBE.

Whittlesford Church. The military graves within the cemetery are where many RAF personnel are buried. *(Author)*

In January 1925, the Meteorological Flight was formed at Duxford. Its role was to undertake high altitude flights daily and gather scientific information and calculations on weather conditions and bring them back for assessment, before being passed on to the Air Ministry. This was at times quite a dangerous task, with pilots often taking off in weather conditions quite unsuitable for normal flying.

One new arrival to Duxford on 28th September 1925 was Geoffrey Tuttle, a young pilot officer under training assigned to 19 Squadron. Tuttle remained with the squadron over the next three years and was eventually promoted. His career in the Royal Air Force was exemplary and he served in nearly every command, bomber, coastal, photo-reconnaissance, training etc, and eventually achieved the rank of air marshal in 1957.

The first day of October in 1925 also saw the formation of the Cambridge University Air Squadron for those young students who wished to undertake flying instruction and perhaps wish to join the RAF as a career. Flying training for the university squadron began in earnest in February 1926, with students being taught and flown in Avro 504s aircraft.

One new officer who arrived at Duxford

during 1927, was a New Zealander, Keith Park, who had served in the Royal Flying Corps in France during the First World War. Park was given command of Treble-One Squadron and would become highly respected by his fellow officers not only in peacetime, but as 11 Group commander during the Battle of Britain and later in the defence of the beleaguered island of Malta, which lay under aerial siege by both the Italian and German air forces.

Treble-One remained at Duxford until it was posted away to Hornchurch in April 1928, while 29 Squadron moved to North Weald. This left 19 as the solitary squadron at Duxford. It continued its peacetime aerobatic flying displays at the air open days and became the Royal Air Force's top display team.

As with all flying, accidents did occur and at Duxford this was no exception. On 28th February 1929, Flight Lieutenant George Bucknall was flying above the aerodrome at approximately 4.30 pm but five minutes later was seen from the ground by witnesses to put his aircraft into a spin at 1,500 feet. He then came out of the spin at 500 feet in the normal manner, but then went into a steeper dive with the engine on full. He managed to flatten the aircraft out of the dive before crashing, but the aircraft caught fire and Bucknall was killed. He was buried at Whittlesford Church cemetery.

His grave has the following inscription:

'This young poet and rare officer, a loving brother, an intimate and abiding friend, much loving and as much beloved was by an accident in the air cut short from his life, to the living of which he had brought so sensitive a zest.' (see photo opposite)

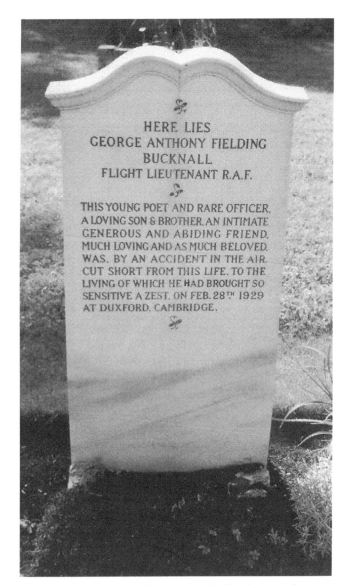

The grave headstone of George Bucknall who served at Duxford during the 1920s. *(Author)*

Opposite page, top: A BE2c biplane is prepared by a ground fitter at Fowlmere aerodrome in 1917, before its pilot will take to the air. *(Courtesy of Martin Sheldrick via Cambridgeshire Collection)*

Opposite page, bottom: An airman stands for the camera with a De Havilland DH6 aircraft at Fowlmere; note the canvas hangar on the right, a type which was in use at this time. *(Courtesy of Martin Sheldrick)*

Top: Another fine mess? A De Havilland DH6 comes to grief crashing beside a thatched cottage situated in Lynch Lane, Fowlmere in September 1918. An airman surveys the scene. *(Courtesy of Martin Sheldrick)*

Bottom: A brand new double Belfast hangar is pictured under construction at Duxford, looking north-east in July 1918. *(A. Raby Collection IWM HU 38689)*

DUXFORD.

No. 35.—Training Depôt Station (Midland Area ; No. 3 Group).

LOCATION.—England, Cambridgeshire, 12 miles south of Cambridge (pop., 56,000). Fowlmere Aerodrome is 4 miles to the west.

Railway Stations :—Whittlesford (G.E. Rly.), 2 miles, and Royston (G.E. Rly.), 7 miles. The main road from Royston to Whittlesford passes the site.

FUNCTION.—A Training Depôt Station (Three Unit), Day Bombing.

ESTABLISHMENT.

Personnel.		Transport.		
Officers	52	Touring Cars		1
Officers under instruction ..	90	Light Tenders		10
N.C.O.'s under instruction..	90	Heavy Tenders		10
W.O.'s and N.C.O.'s above		Motor Cycles		8
the rank of Corporal ..	49	Side Cars		8
Corporals	26	Trailers		5
Rank and File	336			
Foremen	7			
Women	154			
Women (Household) ..	54			
TOTAL (exclusive of Hostel Staff)	858	TOTAL		42

AERODROME.—Maximum dimensions in yards, 1,000 × 1,000. Area, 223 acres, of which 30 acres are occupied by the Technical Buildings. Height above sea level, 100 feet. Soil, light loam. Surface, good, firm, and slightly undulating, with good natural drainage. General surroundings : excellent, being slightly undulating open country with very large fields. The northern half of the belt of trees along the south-western edge (shown in the map) has been felled.

METEOROLOGICAL.—Observations have not yet been made.

TENURE POLICY.—Not at present on the list of permanent stations.

ACCOMMODATION.

Technical Buildings.	Map Reference.
6 Aeroplane Sheds (each 170' × 100')	1
A.R.S. Shed (with Plane Stores) (170' × 100') ..	2
Salvage Shed	—
2 M.T. Sheds	—
Workshop (Wood)	—
Workshop (Metal)	—
Carpenters, 100' × 28'. Sailmakers, 90' × 28'. Dope, 40' × 28'. Machine, 90' × 28'. Engine, 80' × 28'. Coppersmiths, 90' × 15'. Smiths, 90' × 15'.	
Technical Stores	—
Oil Store	—
Petrol Store	—
Instructional Huts—	
General Lecture Hut	—
Gunnery Instruction Hut	—
Gunnery Workshop	—
Photographic Hut	—
Wireless and Bombing Hut	—
Buzzing and Picture-target Hut ..	—
Depôt Offices	—
3 Unit Commanders' Offices	—
Power House	—
Latrines	—
Guard House	—
Compass Platform	—
Machine Gun Range	—
Ammunition Store	—
Officers' Mess	3
4 Officers' Quarters (Staff)	—
3 Officers' Quarters (Pupils)	—
Officers' Baths	—
Officers' Latrines	—
Sergeants' Mess	—
Sergeants' Latrines	—
Sergeants' Baths	—
Regimental Institute	—
Regimental Store	—
4 Men's Huts	—
Men's Baths	—
Men's Latrines and Ablution	—
Reception Station	5
Drying Room	—
Coal Yard	—
Women's Hostel	6

Top: A great aerial view showing Duxford airfield with its old and new style hangars during the beginning of autumn 1918, taken from a DH9 of 123 Squadron during a photographic training exercise. *(IWM Q 11404046)*

Left: An information document giving details of No. 35 Training Depot (Duxford) detailing its function, buildings, establishment etc, as on 1st August 1918. *(IWM HU 41564)*

19

Top: This aerial photograph was initially incorrectly titled as Duxford; it is in fact Fowlmere Aerodrome, taken on 26th April 1918. Fowlmere had been in operational use by 75 (Home Defence) Squadron as a landing ground from early 1917. *(IWM Q 83165)*

Bottom: Members of the Duxford Station Band, taken in 1918, showing a variety of uniforms including the old Royal Flying Corps, the Royal Naval Air Service and the new Royal Air Force uniform, which was still khaki and not blue grey at this time. *(IWM Q 96085)*

Top: Another aerial view of RAF Station Duxford, still under construction on 27th September 1918. Note the temporary hangars and aircraft in foreground. *(IWM Q 96065)*

Bottom: Two training aircraft, a BE2c and a DH6, are damaged after colliding in the middle of Duxford airfield, 1918. *(IWM Q 96076)*

Aircraft engine mechanics with an aero engine on a test-bed frame. *(A. Raby Collection IWM HU40588)*

Could this be the first original flight simulator? Airmen pose for this photograph in one of the wooden huts. There are two cameras, one held by the airman extreme left and one in front of the cut down cockpit. Are they practising for aerial reconnaissance? *(IWM HU 41568)*

Top: Pictured during a break in work time, women of the WRAF smile for the lens. *(A. Raby Collection IWM HU 40568)*

Bottom: Two girls of the Women's Royal Air Force pictured in 1918, driving a motorcycle and sidecar. *(A. Raby Collection IWM Q 86581)*

Top: The Fowlmere Care and Maintenance Party, which was responsible for looking after Duxford as well, shown relaxing at Duxford. *(IWM HU 39406)*

Bottom: Officer pilots are seen wearing a variety of uniforms in this relaxed informal photograph. *(A. Raby Collection IWM HU 40571)*

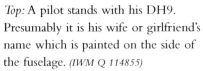

Top: A pilot stands with his DH9. Presumably it is his wife or girlfriend's name which is painted on the side of the fuselage. *(IWM Q 114855)*

Bottom: An RE8 biplane is badly damaged after nosing over during training at Duxford. *(IWM Q 96091)*

Top: A DH9 biplane prepares to take to the air.
(A. Raby Collection IWM HU 41571)

Bottom: Officers and airmen line up in rank for the
Armistice Day Parade in November 1918. *(IWM Q 96081)*

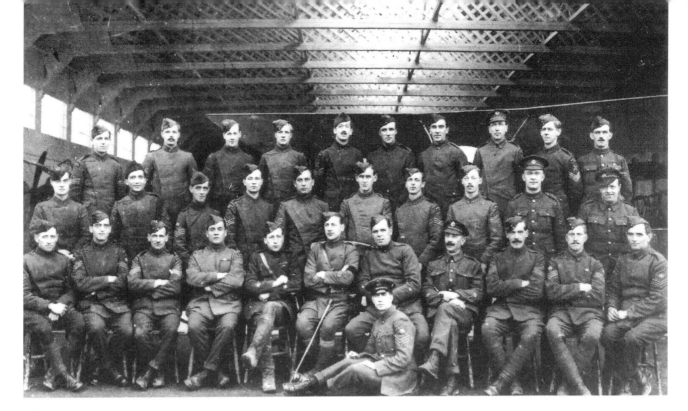

Top: An Armistice Day photograph showing members of the sergeants' mess at Fowlmere airfield, 1918. *(A. Raby Collection IWM HU 41566)*

Bottom: A Handley Page bomber biplane has just landed at Fowlmere and is parked near one of the Belfast hangars, 1919. Note the biplane inside the hangar. *(Courtesy of Martin Sheldrick)*

Top: Various vehicles in the Duxford Motor Transport Bay, including a Hucks Starter, an ambulance and a fire engine, 1919. *(H.A. Hall Collection IWM Q 100274)*

Bottom: An Avro 504 R2443 is pictured during an open fun day at the aerodrome. Below are the fairground attractions including a carousel ride that have been erected on site. *(IWM HU41570)*

Top: WRAFs put all their strength into winning the tug o'war contest during the open fun day. *(IWM Q 114860)*

Bottom: A souvenir postcard produced for Duxford's 2 Training School in 1921 showing the various buildings on site. *(IWM 39316)*

Top: Royal Air Force tradesmen pictured within the carpenter's workshop in 1921. *(J.W. Bayne Collection IWM HU 39317)*

Bottom: A Bristol Fighter being worked on by airframe fitters in one of the hangars, 1921. *(J.W. Bayne Collection IWM HU 39318)*

Top: Five members of the first Non-Commissioned Officers Pilot Course at Duxford, 1921–22, pictured in full flying kit. *(A. Raby Collection IWM HU 41572)*

Bottom: A Sopwith Snipe of 2 Flying Training School comes to grief on an early summer afternoon in 1924. Note the cricket match in the background. *(A/Cdre H.I. Cozens Collection IWM Q 100174)*

Top: An Armstrong Whitworth
Siskin Mk 3 of 111 Squadron, 1925.

(D. Beard Collection IWM Q 103338)

Bottom: 29 Squadron Gloster Grebe aircraft parked up ready for use. In the
foreground is an Avro 504K belonging to the Cambridge University Air
Squadron, 1926. *(D. Beard Collection IWM Q103339)*

Top: Ground personnel of A Flight, 111 Squadron pose for a snap shot with one of their Siskin aircraft, 1926. *(D. Beard Collection IWM Q 103340)*

Bottom: Gloster Grebes of 19 Squadron in one of the hangars in 1927. Flight Sergeant Nash is standing in front of the centre aircraft. *(IWM Q 102726)*

Top: A Gloster Grebe with pilot Ralph Cleland in the cockpit. *(Courtesy of A. P. Pool)*

Bottom: The Gloster Grebe biplane of 29 Squadron in which Sergeant J. Trickett was killed in March 1928, when he crashed near the airfield. *(IWM 41576)*

CHAPTER TWO
SISKINS TO SPITFIRES
1930 – 1939

In September 1931, 19 Squadron exchanged their Siskin aircraft for the Bristol Bulldog IIa.

One Royal Air Force officer who arrived at Duxford in October 1932, was Flying Officer Douglas Stewart Bader. Over a year earlier he had had both legs amputated after a horrific aircraft crash, whilst undertaking a low roll. Having made a remarkable recovery, he now walked into Duxford with the aid of artificial legs. He had been posted to 19 Squadron and hoped that he would be allowed to fly. He was in fact given charge of the motor transport section, but nevertheless was given the chance of an air test and went airborne with instructor Pilot Officer J. Cox, a university air squadron instructor. Although Cox was impressed with Bader's performance, others were to be less sympathetic at the Air Ministry and Bader was retired from the RAF on the grounds of 'ill health'.

Duxford hosted its first Empire Air Display in May 1934, which was extremely successful with 2,385 of the public attending and making £108 for the RAF Benevolent Fund.

In January 1935, 19 Squadron was chosen as the first squadron to be equipped with the RAF's latest biplane fighter, the Gloster Gauntlet with a top speed of 230 mph.

That year Duxford held its most prestigious event to date with the airfield hosting the second part of King George V's Silver Jubilee Review of the Royal Air Force on Saturday 6th July 1935. The first part of the review was to be a ground inspection at Mildenhall by the King and Queen including the Prince of Wales, the Duke of York and Marshals of the RAF, Lord Trenchard and Sir John Salmond. The Secretary of State for Air Sir Phillip Cunliffe-Lister including his deputy, were also on hand, as was the Lord Lieutenant of Cambridge. Foreign dignitaries also attended including two Maharajahs.

On arrival at Duxford the King and his cortège were given lunch at the officers' mess, before being seated at a specially built viewing platform to witness a three-phase flying display which would last thirty minutes. The first display consisted of a flypast by all squadrons, during the second 19 Squadron performed an exhibition of 'squadron drill' and lastly there was a squadron flypast in 'wing' formation which consisted of 182 aircraft.

The King was highly delighted with the review and later wrote to the Secretary of State, asking him to pass on his congratulations to all ranks on the 'magnificent display.'

The routine of peacetime continued unabated apart from the reformation of the First World War 66 Squadron on 20th July 1936 with Squadron Leader V. Croome commanding. The squadron was formed from officers and ground personnel of 19 Squadron's C Flight and was equipped with Gloster Gauntlets. That year a young gentleman by the name of Frank Whittle was studying at Cambridge and could be seen flying regularly with the Cambridge University Air Squadron. Whittle later invented and helped produce the first jet turbine engine which was used in the early jets towards the end of the Second World War. During October, the Meteorological Flight was transferred to Mildenhall in Suffolk after eleven years of service at Duxford.

On Thursday 4th August 1938, Duxford claimed another place in aviation history, when it received the first Supermarine Spitfire Mk 1 to be delivered on charge to an RAF squadron. The aircraft K9789 was flown into the aerodrome by test pilot Jeffrey Quill and was then duly handed over to 19 Squadron, whose commanding officer Squadron Leader Henry Cozens carried out his first flight on the new machine on the 11th.

A few days later 66 Squadron also received its first Spitfire and time was spent by pilots and ground crews in evaluating the aircraft's performance. Both 19 and 66 Squadrons were used for trials during the following months to see if any improvements could be made to the aircraft, for example, the two-bladed fixed-pitch propeller, radio equipment and the retractable undercarriage system were all added. Two of the new aircraft were to be put through 400-hour trials as soon as possible, then to report the findings. Throughout the autumn the squadrons received the new aircraft at the rate of only one per week and it would not be until the end of the year, before they would be up to a squadron strength of sixteen aircraft.

Accidents were inevitable and on 3rd November the first Spitfire was written off, when a faulty axle stub sheared off during landing, its pilot Gordon Sinclair fortunately being unhurt, even when the aircraft K9792 turned over.

The aerodrome was brought to a state of emergency on 26th September 1938, when along with other RAF stations, it was placed on a two-hour alert and procedures were put on a war footing, as things began to look serious with the threat of Hitler's Nazi Germany seeking to expand its borders to take over the Sudeten area of Czechoslovakia. This crisis which brought Europe to the brink of war was only averted by an agreement, later made famous by Prime Minister Chamberlain's 'piece of paper' which brokered a deal by the leaders of France and Britain to allow Germany to take over the Sudetenland part of Czechoslovakia and ostensibly gained Hitler's agreement to halt any further plans of expansion. Chamberlain returned from Munich to Croydon airport proclaiming 'Peace in our Time.' With the emergency over, sadly only temporarily, the aerodrome returned to normal duties within a few weeks and on 8th October, a flight of 19 Squadron flew over to display their Spitfires at Marshall's airfield in Cambridge.

By the beginning of 1939, the squadrons at Duxford had completed many hours of flying and gained invaluable experience on their

Spitfire aircraft. This was to be the real value of the Munich agreement. On 4th May Duxford became the focal point for correspondents and photographers of the Press who were invited to see a demonstration by 19 Squadron with their Spitfires including a simulated air raid and pilots scrambling to their aircraft.

August saw the arrival of Royal Auxiliary Air Squadron 611 (West Lancashire) to the station for summer camp; they too were equipped with Spitfires.

A state of emergency was again sounded at aerodromes around the United Kingdom that month as the storm clouds of war seemed to draw ever nearer with Adolf Hitler's reluctance to heed the warning to stop the expansion of the Third Reich's empire.

At Duxford on 24th August, 19, 66 and 611 Squadrons were all called to thirty minutes readiness and all personnel on leave were recalled. Ground personnel prepared the aerodrome for protection by filling sand bags and constructing dispersal pens. Security was increased around the station on the 29th and reservists were called up.

German troops crossed over the border to attack Poland on 1st September and an ultimatum was sent to Hitler to withdraw his armies or a state of war would be declared by Britain and France. By 3rd September, no communication from Hitler had been received, so at 11.15 am that morning Prime Minister Neville Chamberlain spoke to the nation to tell them Britain was now at war. At Duxford, the communications office received Signal A34 which confirmed that war had indeed broken out with Germany.

Preparation for the conflict continued apace with modification to barracks and hangars with camouflage paint being applied, as well as further accommodation being installed for the influx of more WAAF personnel.

The aerodrome soon suffered its first flying casualty of the war although it was not in combat. On the night of 6th September, Pilot Officer Douglas George Paton of 66 Squadron was undertaking a night-flying exercise, when his aircraft, Spitfire K9968 stalled at 300 feet and crashed into trees on the south side of the aerodrome and burst into flames.

A state of high tension was felt around the aerodrome during the next few days, with the expectancy of German bombers arriving at any time, any minute to attack the aerodrome. In reality this was complete nonsense as no German aircraft had the range to cover that distance from Poland or Germany. And soon the fear of invasion relinquished and it was back to the daily routine, so much so that by Christmas of that year, with so little action against the Germans the conflict became known as 'The Phoney War.'

On 5th October, 222 Squadron was reformed at Duxford under the leadership of Squadron Leader Herbert Waldemar Mermagan to fly Bristol Blenheims and operate as a night-fighter unit.

Top: An aerial view showing one of the hangars and the surrounding buildings in 1932. *(Diane Crain Collection)*

Bottom: An Armstrong–Whitworth Atlas which was used for research into airflow characteristics by the Cambridge University Department of Aeronautics in 1932. The work was supported by the Armourers and Braziers Company, whose badge was carried on the fuselage. *(IWM HU 18156)*

Top left: Avro Tutor biplanes of the Cambridge University Air Squadron at the aerodrome in 1932. *(Diane Crain Collection)*

Top right: A Westland Wallace of the Meteorological Flight is given a helping hand by ground crew, while taxiing. *(Diane Crain Collection)*

Bottom: A Crossley Tender truck with RAF personnel; note the aircraft being towed behind. *(Diane Crain Collection)*

Top: This rare picture of Douglas Bader was taken during a Duxford Station Treasure Hunt held on 8th March 1933 just sixteen months after Bader had lost both his legs in a flying accident. Three weeks after it was taken he was retired by the Royal Air Force on the grounds of ill-health. Thelma Edwards, later to become his wife, is also in the photograph. *(Diane Crain Collection)*

Bottom: A Bristol Bulldog of 19 Squadron. This type replaced the Siskin in 1932 before giving way to the Gloster Gauntlet in 1935. *(IWM HU 18163)*

Top: The Royal Air Force Duxford Station Band, pictured on the parade ground in October 1934.
(IWM HU 49741)

Bottom: An excellent view of Duxford's hangars with doors open, during the 1935 Empire Air display. Three Avro Tutors of the Cambridge University Air Squadron are pictured and a Handley Page Heyford biplane bomber is parked up outside the first hangar on the left. *(IWM HU 48146)*

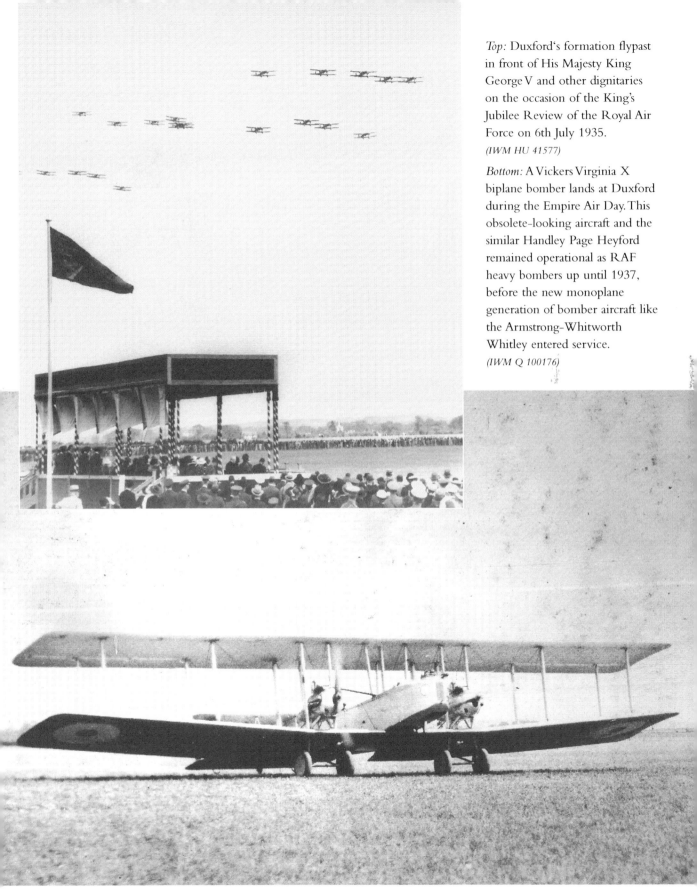

Top: Duxford's formation flypast in front of His Majesty King George V and other dignitaries on the occasion of the King's Jubilee Review of the Royal Air Force on 6th July 1935.
(IWM HU 41577)

Bottom: A Vickers Virginia X biplane bomber lands at Duxford during the Empire Air Day. This obsolete-looking aircraft and the similar Handley Page Heyford remained operational as RAF heavy bombers up until 1937, before the new monoplane generation of bomber aircraft like the Armstrong-Whitworth Whitley entered service.
(IWM Q 100176)

Top: A Gloster Gauntlet II as used by the Meteorogical Flight based at Duxford. The Flight had been formed at Duxford in January 1925 and carried this important work until they moved to Mildenhall in 1936.

(Copyright Air Historical Branch)

Bottom: Officers of 19 (F) Squadron pictured in front of one of their Gloster Gauntlet biplanes in 1936. Officers known are front row left to right: F/O B.G. Morris, F/O J. R. MacLachlan and F/Lt H. Broadhurst. Back row 2nd right is P/O J.A. Kent. *(Author via H. Broadhurst)*

Top: 19 Squadron's famous aerobatic team of MacLachlan, Broadhurst and Morris, wearing their pre-war white flying overalls. Their performance of aerobatics thrilled those who attended the 1936 Hendon Air Pageant, when they proceeded to undertake the aerial manoeuvres whilst all three aircraft were connected together by cords to the wings.
(Author via H. Broadhurst)

Bottom: The de Bryne Snark monoplane which was built by Aero Research Limited of Duxford and used for research flying by the Cambridge University Department of Aeronautics in 1936-37.
(IWM HU 18158)

Top: British aviation designer Reginald Joseph Mitchell's masterpiece, the prototype Supermarine Spitfire K5054 pictured at Duxford in 1937. *(Author via Deere Collection)*

Bottom: The front exterior of Building 6, the airmens' dining hall and canteen viewed from the south-west, April 1938. *(Leslie Bridges IWM HU 65353)*

Top: Building 7, then 19 (F) Squadron's airmens' domestic block, viewed from the south-west, April 1938. *(Leslie Bridges IWM 65351)*

Bottom: Officers of 19 Squadron pose for a group photograph in front of one of their Gloster Gauntlet aircraft during Duxford's Empire Air Day in 1938. *(IWM HU 27835)*

Top: Instructors in flying overalls talking with ground staff seated at a trestle table outside the flight hut (the Watch Office) in front of Hangar 78, during the University of London Air Squadron summer camp in 1938. *(IWM HU 65341)*

Bottom: A group of eager University of London Air Squadron members await their turn, dressed in white flying overalls, outside the flight hut. *(IWM HU 65344)*

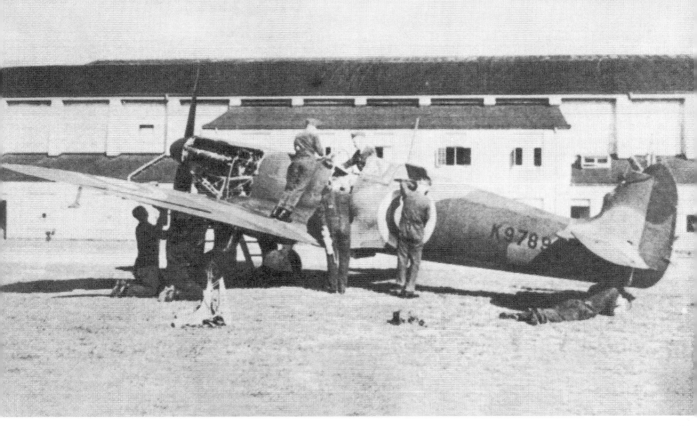

Top: A new era was heralded in at Duxford on 4th August 1938, when Spitfire K9789 arrived and was put on charge to 19 Squadron, the first RAF squadron to receive the new fighter aircraft. The squadron's commander Henry Cozens made his first flight in the aircraft on 11th August.

Bottom: Spitfire K9790 was the second aircraft to arrive at Duxford and was used for intensive flight trials by 66 Squadron. *(IWM HU 41583)*

49

Top: Spitfires of 19 Squadron are lined up for a special press day on 4th May 1939. This image amply demonstrates the two-bladed propeller being used at this stage of the aircraft's development. *(IWM HU 48148)*

Right: Two ground-crewmen of 66 Squadron pictured with their aircraft, 1939. *(IWM HU 35587)*

Opposite page: The first official photographs of the first RAF squadron to receive the new Supermarine Spitfire were taken when 19 Squadron took to the air on 31st October 1938. This photograph was taken from a Blenheim of 114 Squadron. Leading the formation was Squadron Leader Henry Cozens pictured in foreground. *(IWM CH 19)*

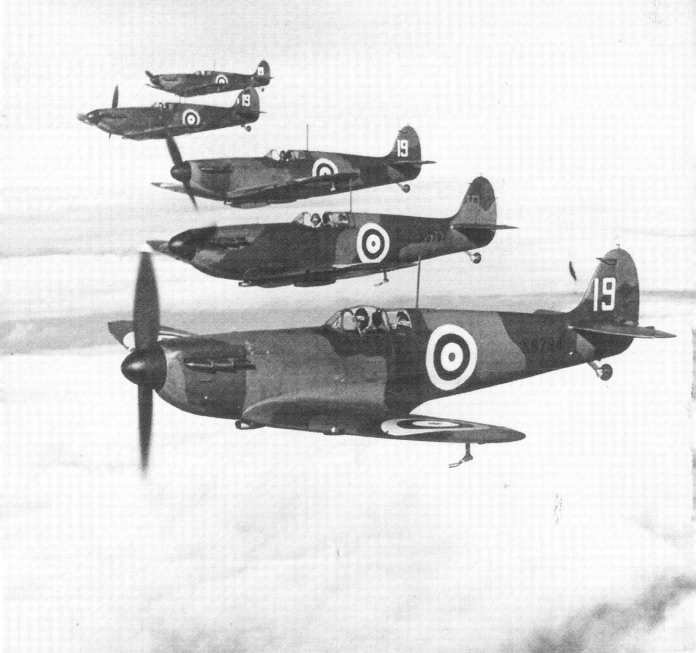

One of the
machine-gun butts
located near the
American Air
Museum as it is
today and (inset)
a Spitfire at the
butts in early 1939.
(IWM HU 55027)

Top: 19 Squadron at dispersal on the south side of the aerodrome after the outbreak of war in September 1939.
Left to right standing:
P/O Lyne, P/O Watson, S/Ldr Cozens, unidentified, P/O Sinclair, F/O Coward, F/Sgt Steere, unidentified, F/Sgt Unwin, Sgt Potter, unidentified. Middle seated: F/O Petre, F/Lt Withall, P/O Llewellyn, P/O Ball. Front row: F/O Pace, P/O Brinsden, F/Lt Banham, F/Lt Clouston, unidentified. *(IWM HU 27858)*

Bottom: Armourers of 66 Squadron at their dispersal manage to find time for a quick snap shot. *(IWM 35588)*

This page: A Spitfire of 19 Squadron with engine exhaust stacks smoking, prepares to taxi out from its dispersal, early September 1939.
(A/Cdr H.I. Cozens IWM HU 27859)

Opposite page, top: Pilots of 611 'West Lancashire' Auxiliary Squadron in October 1939. Left to right standing: P/O P.S. Pollard, P/O Sidney Bazeley, P/O Colin Macfie, F/Lt Jack Leather, P/O Denis Adams, F/O Little, F/O Hamilton, S/Ldr James McComb, P/O Peter Brown, P/O Mitchell and F/O Crompton,. Kneeling: F/O Douglas Watkins. *(IWM HU 45436)*

Opposite page, bottom: Members of 117 Light Anti-Aircraft Artillery Regiment, Royal Artillery based at the airfield during winter 1939/40. *(IWM HU 56414)*

CHAPTER THREE
TO DEFEND AND ATTACK
1940 – 1942

The winter of 1939-40 was an extremely cold one, with temperatures dropping well below zero. Duxford was snowed in and very few flying days were accounted for during January. 611 Squadron had departed in October 1939, leaving 19 and 66 Squadrons, who were given the tasks of evaluating new equipment for various Air Ministry departments.

66 undertook trials of the new VHF radio systems, while 19 was given the job of having some of their aircraft fitted with newly developed 20-millimetre Hispano cannons into the wings of their aircraft. The new armament would give the aircraft harder hitting destructive power, but the trials showed that the weapons frequently jammed.

The first action of the war as far as Duxford's squadrons were concerned, took place on 11th January 1940, when 66, which had been sent to the airfield at Horsham St. Faith near Norwich, receiving information from the operations room, were vectored on to an enemy aircraft. A Heinkel He111 bomber had attacked a fishing trawler and while fleeing from the scene was in return intercepted and attacked by three Spitfires. The pilots reported damaging the port

engine of the Heinkel before losing it in cloud. It was later reported that the German aircraft had crashed in Denmark.

222 Squadron had reformed and was flying Blenheims when it received the news on 31st January that Pilot Officer David Giles Maynard had been killed when his aircraft L6614 had crashed into a hill near Royston, Hertfordshire. The cause of the accident was due to bad visibility. He lies buried at Whittlesford Cemetery grave B3. 222 Squadron lost two further pilots due to accidents during the next following weeks. Pilot Officer Arthur Delamore on 18th February and Pilot Officer James Griffiths aged twenty-one on 28th March, who while flying a Miles Magister trainer, crashed into the ground from on top of a loop at Fowlmere during a landing and local flying exercise.

19 Squadron saw the return of Douglas Bader to their ranks on 7th February after he had successfully passed a flying refresher course and persistently argued that he was medically fit, apart from his artificial legs. Bader was excited on rejoining the squadron and being able to fly the much vaunted and prized

Spitfire. But his inexperience on the type would see him involved in another accident, on 31st March, when he took off without enough forward air-speed to get airborne, and hit the boundary fence at the end of the flight-path. Fortunately Bader was uninjured, but the aircraft was a total write-off.

Duxford had received a change in station commander on 21st March, when Group Captain A. B. 'Woody' Woodhall replaced Wing Commander Lester. Woodhall, an ex First World War pilot, was an energetic character and wore a monocle. During March, 222 Squadron was re-equipped with Spitfires and in April 19 Squadron moved to Horsham St. Faith on a semi-basis.

Hitler's armies started their next strategic phase of the war, when they advanced into Holland and Belgium on 10th May 1940, breaking through the Ardennes with their Panzer tanks and infantry. Their use of the Luftwaffe as a spearhead to knock out enemy positions became known as Blitzkrieg. Very soon the Allied armies in France were pushed back in disarray, due to the fast moving German tactics. By 23rd May, the situation over in France was so bad that a plan to save and evacuate the British Expeditionary Force Army was put into action. The BEF was retreating back towards the French port of Dunkirk, but fighting every inch of ground against the surrounding German forces. While the Royal Navy was tasked with the job of rescuing as many soldiers from the beaches and port of Dunkirk, the Royal Air Force's role was that of stopping the Luftwaffe in bombing and strafing the beaches below.

Duxford's fighter squadrons were duly ordered south to cover the withdrawal. 19 was sent to operate from RAF Hornchurch, while 264 Squadron went down to Manston in Kent. From 26th until 1st June, the Duxford pilots flew dozens of sorties across the Channel against the Luftwaffe. 19 Squadron claimed twenty-eight destroyed and ten damaged, 264 claimed forty-nine destroyed and nine damaged. Several pilots were killed or became prisoners of war.

Britain now had its back firmly against the wall and with Europe occupied by the Germans, the citizens of the British Isles waited for Hitler's next move. Would it be invasion?

On 25th June, 19 Squadron was sent to operate from Fowlmere, which by now had been brought back into operation as a satellite airfield. However, it still only had the basic requirements with which to house a squadron, including canvas tents, Nissen huts and mobile communications.

On 10th July, according to the records, the Battle of Britain officially began. On that same day at Duxford, a new squadron was formed made up of Czechoslovakian pilots, with RAF Squadron commanders; this was 310 (Czech) Squadron. Many of the pilots had already fought the Germans in the skies of their own country and had fought across Europe before escaping to Britain to carry on the fight against the Nazis. One of the main problems at first was that of language not only on the ground, but in the air, but this was soon resolved with Czech translators teaching them basic RAF terminology. The squadron was equipped with Hawker Hurricane Mk 1s.

On 13th July 19 Squadron suffered the loss of one of its pilots, when Sergeant Raymond Birch in Spitfire R6688 stalled his Spitfire in a steep turn during a dogfight practice at 7.00 pm near the airfield. The aircraft crashed to the ground and burned killing the young pilot. He was buried at Whittlesford Church Cemetery.

The first phase of the Battle of Britain had now begun with the Germans launching raids against convoy shipping and attacks against coastal ports and military installations. While this was being mainly dealt with by squadrons in 11 Group based in the south-east, the squadrons in 12 Group waited for the call to action. Patrols were flown throughout this period, but interception with the enemy was minimal.

On 6th August Dr Eduard Benes, President of the Czechoslovak National Committee in exile, arrived at Duxford at 4.00 pm accompanied by General Slezak, Officer Commanding the Czech Air Force, Dr Smutny, the Councillor of the Czech Legation and Air Vice-Marshal Trafford Leigh-Mallory, Air Officer Commanding 12 Group. Benes inspected a guard of honour supplied by the station defence and addressed the Czech squadron before taking the salute at a march past. The party then witnessed some formation flying and aerobatics by Hurricanes of 310 Squadron.

The Czech squadron became operational on 17th August 1940. At the end of August another Czech squadron, 312, was formed at the aerodrome; this too flew Hurricanes.

The war was brought closer to home on the night of 28th August, when at 10.00 pm a Dornier 17 bomber was shot down by a Bofors gun of 243 Battery, 78th Anti-aircraft Regiment. Two days later 242 Canadian Squadron arrived, led by Squadron Leader Douglas Bader. Bader had been given command of 242 in June, the squadron having suffered heavy casualties during the French campaign. He had brought them back to a fully operational squadron and had some success

against the enemy flying from RAF Coltishall in Norfolk. The 12 Group commander Trafford Leigh-Mallory, impressed with Bader's tactics and 242 Squadron's claims of aircraft destroyed, decided to operate them from Duxford. On this day, the Duxford squadrons were ordered to readiness as signals from 11 Group asked 12 Group to send its squadrons to cover the 11 Group airfields against attack from any German formations that got through the fighter defence. 242 Squadron led by Bader was ordered to patrol North Weald aerodrome, and remain in this vicinity against possible attack from the Luftwaffe.

Bader, impatient with waiting for the enemy to appear, changed course to seek out any German formations. Fortunately, he came across a group of fifty-plus Dornier Do17s and Messerschmitt Bf110s west of Enfield and dived down to attack. On return to the aerodrome, the squadron claimed twelve enemy aircraft destroyed. In reality it was later found that only four enemy aircraft had been shot down.

Air Vice-Marshal Leigh-Mallory was delighted with the figures of enemy claimed shot down by 242 Squadron and was swayed by Bader's argument that if he'd had more aircraft, then they could have shot down more. He decided that Bader should have more squadrons at hand and lead a formation of three squadrons as a 'Big Wing.' 19, 242 and 310 Squadrons would make up the wing. This tactical policy would cause friction and disagreement between the Fighter Command group commanders of 10 and 11 Group, whose tactics thus far had proven successful against the Germans. Air Vice-Marshal Keith Park claimed that there was no need for a Big Wing and that it took too much time to form up a large formation of aircraft, up to twenty minutes, while in the meantime the enemy was about to or had already attacked their targets.

The only Battle of Britain pilot casualty buried nearby is that of Pilot Officer Raymond Andre Charles Aeberhardt who flew with 19 Squadron operating out of Fowlmere. On Saturday 31st August 1940, Aeberhardt in Spitfire R6912 was sent off to intercept a large formation of enemy aircraft who were heading to bomb RAF Debden and Duxford. On entering combat with the German formation, his Spitfire was damaged by gunfire to the hydraulic system which operated the flaps and undercarriage systems. On return to Fowlmere at 8.50 am Aeberhardt's aircraft was seen to touch down, but it then immediately somersaulted onto its back and caught fire. He died trapped in his cockpit aged just nineteen years. He was buried with full military honours at Whittlesford, grave A7, where he lies still.

Above: Pilot Officer Raymond Aeberhardt from Walton-on-Thames. *(After the Battle Collection)*

Left: The grave of Battle of Britain casualty Raymond Aeberhardt killed on 31st August 1940. *(Author)*

On 5th September, the station played host, ironically, to eight Japanese journalists, who were shown around various areas of interest, from which they could write an article. Within fifteen months, Britain would be at war with the Japanese Empire.

On 14th September, the Duxford Big Wing was increased by two further squadrons making a total of five – 19, 242, 302, 310 and 611. The following day, 15th September now commemorated as Battle of Britain Day, saw the Big Wing scramble during the late morning as a major raid against London was picked up by Radio Direction Finders (RDF, later known as Radar).

As the Wing arrived over London battle had already been met by squadrons or

individual aircraft from 11 Group. But the sight of a large formation of sixty RAF fighters must have completely demoralised the German aircrews, whose intelligence had told them that the RAF was down to its last few fighter aircraft. The Duxford Wing claimed forty-four enemy aircraft destroyed with eight probables. This with Fighter Command's overall claims of 185 aircraft destroyed, looked very impressive, but in reality the RAF only destroyed fifty-six aircraft that day. The significance of the Wing was seen more as a morale booster than as an effective tactical weapon, indeed their over-claiming ratio was 3 to 1.

The Wing was sent off on three separate occasions on 18th September, but on only the third patrol did they encounter enemy action between London and Thameshaven. During the combat, four Junkers Ju88s were shot down, two each by 19 Squadron and 302 Squadron. During the rest of September and the whole of October, the Wing flew nineteen patrols, but made no interceptions.

By the end of October, the Germans were making little effort on daylight raids; instead they now concentrated their main thrust on night raids against London and other large cities. On 30th November 242 left Duxford and returned to Coltishall, leaving 19 and 310 Squadrons to maintain defensive patrols.

The Czech President Dr Benes returned again on 14th December, to present awards to Wing Commander A.B. Woodhall and several British and Czech personnel of 310 Squadron. Benes was accompanied by General Ingr, the Minister of National Defence.

By the end of 1940, with the threat of invasion now seemly at an end for at least that year, the squadrons at Duxford carried on their routine procedure of protecting the Midlands and the shipping convoys around the north-east. A new unit arrived at the aerodrome during that December, whose main task was to evaluate new aircraft and technical systems; it was known as the Air Fighting Development Unit.

In January 1941, another new unit arrived and operated from Duxford. This was the Air Gun Mounting Establishment, which specialized in the development of new armament and its effectiveness after being installed in fighter or fighter-bomber aircraft.

On 14th January, His Majesty King George VI arrived for an award ceremony. After lunching at the officers' mess, the King presented decorations to pilots and personnel witnessed by Air Vice-Marshal Saul, the officer commanding 12 Group.

Duxford received its first enemy bombing raid of the war on the night of 25th February 1941 at 9.55 pm, when the flare-path was hit by eight bombs. A second attack occurred at 10.10 pm, this time with incendiaries and one high explosive bomb was dropped and hit the dispersal of 310 Squadron causing a fuel bowser to explode, killing two Czech airmen, aircraftsmen Benedikt Pohner and Jaroslav Zavadil and injuring five others. The Czech airmen are buried in the cemetery at Whittlesford Church. They are the only victims that Duxford suffered due to enemy bombing during the entire war.

In March the AGME received a new chief technical officer who was also chief test pilot; this was Squadron Leader John Gray Munro who had served with 263 at Grangemouth flying Westland Whirlwinds. During that month a Bristol Beaufighter R2055 arrived from Filton for installation of two 40-millimetre cannon and to conduct recoil testing.

On 18th June 787 Squadron Fleet Air Arm, a Naval Air Fighting Development Unit from Yeovilton, arrived and set up camp on the south-east corner of Duxford. Its role was to carry out trials on improving naval aircraft which included a Blackburn Skua and Fairey Albacore.

On 26th June 1941, 310 Squadron departed Duxford for its new posting to operate from Martlesham Heath in Suffolk; its position was taken in return by Martlesham's 56 'Punjab' Squadron flying Hurricane Mk IIbs. Its commanding officer was Squadron Leader Peter 'Prosser' Hanks DFC, who had previously been with 1 Squadron during the Battle of France and was an extremely experienced pilot. During the next few months, 56 Squadron would undertake offensive sweeps across the Channel. The squadron would fly down to West Malling aerodrome in Kent, to use as an advanced base, before heading out to seek out the enemy.

August 15th saw the departure of Duxford's longest serving squadron, 19, which after giving eighteen years of service left to undertake operations at Matlask, the satellite airfield for RAF Coltishall, The aerodrome received an American volunteer squadron, 133 'Eagle' Squadron, on the same day. American pilots who had decided to join the Royal Air Force during Britain's hour of need in 1940, while the United States had remained neutral, had now formed into three Eagle Squadrons, 71, 121 and 133. The squadron's stay was brief and it moved to operate from Fowlmere in October.

19 Squadron was replaced by 601 'County of London' Squadron which had become the only RAF squadron to be equipped with the American-built Bell P39 Airacobra fighter aircraft. The squadron converted to these aircraft at Duxford. The aircraft was notable for its tricycle undercarriage, the engine being fixed behind the pilot's cockpit, with the propeller shaft running between the pilot's legs and a canopy which opened and looked similar to a car door. The aircraft's armament included a 20-millimetre cannon that fired through the propeller spinner. Trials of the P39 found the aircraft wanting in various categories of performance and the first two flights resulted in the pilots having to force land. Early teething problems included the cockpit being filled with fumes when the nose-mounted cannon was used. It was found however to perform better at low-level altitude. The only operation carried out by the squadron using the P39 took place on 9th October, when a section was sent down to Manston to take part in a raid on the French coast.

Another unit based here at this time was 74 (Signals) Wing who carried out a variety of work including coastal radar calibration. They flew a selection of aircraft including a Blenheim MkIV, Hornet Moths and the strangest of aircraft which looked like the a cross between an aeroplane and an early helicopter; the Cierva C30A Autogiro.

A new sight and sound was heard on the aerodrome on 31st August 1941, when a new secret RAF fighter landed. This was the Hawker Typhoon serial No.R7580, armed with four 20-millimetre cannon, with a speed of 400 mph and larger than any other previous fighter aircraft designed for the Royal Air Force. Once the aircraft had landed it was taxied to the Air Fighting Development Unit hangar to undergo various trials.

On 11th September, 56 Squadron was given its first Typhoon aircraft R7583 to familiarize the pilots and ground crews with the workings of the new aircraft and also carry out further flying testing.

On 12th September, local residents could be mistaken for thinking that the Germans were invading the area around Duxford, when a selection of various captured Luftwaffe aircraft descended on the aerodrome for evaluation and demonstration. These were flown in by 1426 (enemy aircraft circus) Flight and consisted of two Junkers Ju88s, a Heinkel He111, and a Messerschmitt Bf 110 and 109. All the aircraft had RAF roundels painted on the wings and fuselage.

Another aircraft flown into Duxford for trials by the Air Fighting Development Unit was a Messerschmitt Bf 109F. This aircraft had been captured intact on 11th July 1941, after its pilot Hauptmann Rolf Pingel of I/JG26 squadron had been attacked by Spitfires over the Channel and forced to land his aircraft at St Margaret's Bay in Kent. It had suffered little damage and after initially being taken to the Royal Aircraft Establishment at Farnborough it was sent to Duxford on 11th October 1941. On 20th, wearing RAF roundels it was taken on a test flight to carry out a comparative speed trial against a Spitfire. During the test, the aircraft was seen to dive at an angle of 50 degrees which steepened further into a vertical dive before crashing into the ground at Green Lane, Fowlmere. The official Air Ministry report for the incident recorded that the Polish pilot, Flying Officer Marian Skalski, had suffered from carbon monoxide poisoning and had passed out. Skalski was buried with full military honours at Whittlesford Church cemetery.

Problems with the Typhoon continued and especially on 1st November when an officer of 56 Squadron dived straight into the ground from a height of three thousand feet showing

no effort to pull the aircraft out of its dive. It was discovered that he had also been poisoned by carbon monoxide gas and all the aircraft were grounded until the problem was solved.

The squadron received a new commanding officer on 22nd December, when Squadron Leader Hugh 'Cocky' Dundas DFC arrived. Dundas had fought previously in the Battle of Britain with 616 Squadron and then with Douglas Bader's 'Tangmere Wing.'

On 4th January 1942, 601 Squadron moved out of Duxford to operate from Acaster Malbis, the satellite airfield for RAF Church Fenton.

January 30th saw the arrival of 266 Squadron from Wittering under the command of Squadron Leader C.L. Green to operate the new Hawker Typhoon aircraft alongside 56 Squadron. Both squadrons found the new aircraft had many problems yet to be ironed out, before they would become fully operational. Pilots suffered from engine cut-outs in flight which resulted in many crashes and on landings the tail-wheel would collapse.

On 12th February, the aerodrome received a new station commander, Group Captain John Grandy, who had previously led 249 Squadron during the Battle of Britain at North Weald, and had come from being wing commander flying at Coltishall.

Many types of new aircraft were being tested by the Air Fighting Development and Air Gun Mounting Establishment at this time. The De Havilland Mosquito whose main construction was entirely of wood was going through trials secretly at Duxford as was the early Mk 1 North American Mustang. Other aircraft were being tried and tested with new more hard hitting armament; these included a Vickers Wellington with a massive 40-millimetre cannon which was positioned in the tail of the aircraft. In March, 609 Squadron arrived and it too began to convert to the Hawker Typhoon.

During July 1942, Duxford's satellite at Fowlmere became the unlikely participant for the making of a contemporary war movie. Two Cities Film Company were producing a film titled 'In Which We Serve' starring Noel Coward as Captain Kinross, John Mills as Ordinary Seaman 'Shorty Blake', and also starring Bernard Miles and a 17-year-old Richard Attenborough. The film was co-directed by Coward and David Lean in his debut directing role. It told the factious story of the officers and men of Royal Naval destroyer HMS *Torrin* which was loosely based on Lord Mountbatten's ship, the HMS *Kelly* and her sinking. One sequence in the film called for the ship to be attacked from the air by a German bomber. A dummy superstructure of the destroyer was built and set up at Fowlmere and the German bomber was supplied by Duxford's Enemy Aircraft Flight using a captured Junkers Ju88. One can but imagine the local people's response to seeing a low-flying German aircraft diving on the Fowlmere airfield throughout the filming.

On 1st September a fourth squadron, 181 was formed at Duxford, its commanding officer being Squadron Leader Denis Crowley-Milling, who had flown with the Duxford Big Wing in 242 Squadron during 1940.

The squadron suffered a fatality on 27th September, when Hawker Typhoon R7676 flown by Flight Lieutenant Gordon Lindsell spun into the ground at the aerodrome during a practice flight.

The United States of America entered the war on 7th December 1941, having been attacked by the Japanese at Pearl Harbor, and when Hitler declared war on the US, she was now totally committed to waging war against the Germans as well as the Japanese. Consequently, on 12th August 1942 Duxford saw the arrival of the first American pilot, when 1st Lieutenant J.A Glenn flying a Lockheed P.38 Lighting fighter flew in from Goxhill conducting tactical trials. Glenn belonged to the 1st Fighter Group United States Army Air Corps.

More American airmen arrived at Duxford on 1st October 1942, when three fighter squadrons, the 345th, 346th and 347th of 350th Fighter Group under the command of Major R. F. Klocko flew in. Only the 345th Squadron remained at Duxford with its Group Headquarters. The 346th was sent on detachment to Coltishall, the 347th to Snailwell. Spitfires and a few Airacobras were being used while the fighter group began the build-up to operational strength.

In December 1942, the RAF's 169 Squadron arrived from Clifton flying the North American Mustang, to start operations comprised of ground-attack and shipping reconnaissance missions.

Opposite page, top: The figure of Sergeant F.N. Robertson stands alone against the backdrop of 66 Squadron Spitfires at the snow-covered fighter airfield in February 1940. *(IWM HU 2377)*

Left: Pilot Officer Peter Howard-Williams pictured with his Spitfire coded QU-R at the satellite airfield of Fowlmere. *(Author via P. Burgess)*

Below: Group photo of officers and airmen of 66 Squadron pictured outside one of the camouflaged Belfast hangars. *(IWM HU 35048)*

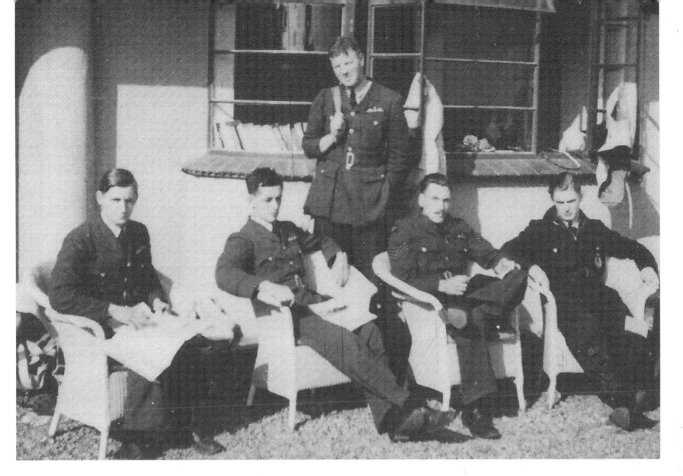

Opposite page: A Bristol Blenheim of 222 Squadron prepares to take off on another training flight. *(Courtesy of A. P. Pool)*

Top: Pilots of 19 Squadron relax outside the crew room in Lloyd Loom chairs, May 1940. Left to right: Pilot Officer Michael Lyne, Pilot Officer Peter Howard-Williams, Flying Officer Geoffrey Matheson (standing), Sergeant Jack Potter, Pilot Officer Peter Watson. Matheson was killed in 1943, while flying Mosquitoes with 418 Squadron; Potter became a prisoner of war after being shot down into the Channel on 15th September 1940 and Watson was killed on 19 Squadron's first operation over Dunkirk on 26th May 1940. *(Author via P. Burgess)*

Bottom: Station commander Wing Commander A. B. 'Woody' Woodhall pictured in his office at Duxford. Woodhall took over command from Wing Commander H.L.P. Lester on 21st March 1940 and became a highly respected and likeable commander with pilots and ground personnel. *(IWM CH 1386)*

Top: Officers of 19 Squadron pictured at Fowlmere, July/August 1940. Left to right: Padre Mayfield, Squadron Leader Brian Lane, Sergeant Bernard Jennings, Flying Officer Noel Brinsden and Pilot Officer Wallace Cunningham. *(Author via P. Burgess)*

Bottom: Men of 264 Squadron in front of a Boulton Paul Defiant. 264 had great success over Dunkirk on 29th May 1940, flying from Duxford, with thirty enemy aircraft claimed for the loss of one air gunner. Back row left to right: P/O Hickman; F/L N.G. Cooke; S/L P. Hunter; P/O M.H. Young; P/O Hackwood; P/O E. Barnwell; P/O Thomas; P/O Whitley. Front row left to right: Sergeant Thorn; P/O Desmond Kay; Sergeant Lauder; P/O Stokes.

Top: Three pilots of 19 Squadron pose for an official photographer after being awarded decorations for gallantry and devotion to duty during operations over France during the Dunkirk evacuation. Left to right: Flight Sergeant Harry Steere DFM, and Squadron Leader Brian Lane and Flight Lieutenant Wilfred Clouston who were both awarded the DFC. The pilot far right is a Czech who won the Croix de Guerre while serving with the French Air Force in France. *(IWM CH 1325)*

Bottom: Flight Sergeant George 'Grumpy' Unwin DFM with his playful Alsation Flash at Fowlmere. Unwin had been with the squadron since 1936 and ended the war as an ace with a total of thirteen and two shared destroyed. He remained in the service after the war and retired as wing commander in 1961. *(IWM HU 58946)*

Top: RAF officers of 310 Squadron assemble to listen to Dr Eduard Benes on a visit by the Czech President in exile to the airfield on 6th August 1940.
(IWM HU 40535)

Bottom: Ground crew hard at work getting another Spitfire ready for its next sortie. Mechanics, riggers, armourers and wireless engineers etc, all got on with their important tasks with a professionalism much admired and appreciated by the pilots. *(IWM 1359)*

Top: Having been previously with 19 and 222 Squadrons at Duxford, Flight Lieutenant Douglas Bader was given the job of acting squadron leader of 242 Canadian Squadron at Coltishall in July 1940, and by August was operating from Duxford.

After persuading his 12 Group commander Air Vice-Marshal Trafford Leigh-Mallory to introduce a change of aerial tactics with the use of greater numbers of squadrons in large formations (The Big Wing), he envisaged that larger numbers of enemy aircraft could be shot down. Bader led a wing formation of five squadrons from Duxford, but controversy soon raged between those in favour and those opposed to the new tactics. *(Author)*

Bottom: Czech armourers of 310 Squadron work on Hurricane P3143, the aircraft flown by Flight Lieutenant Gordon Sinclair. *(Author via P. Burgess)*

A fine study of Czech Sergeant Pilot Raimund Puda standing by his Hurricane.

Puda joined 310 Squadron at Duxford on 16th August 1940, after fighting his way through France as a pilot with the French Air Force, following the German invasion of Czechoslovakia in March 1939. With the fall of France in June 1940, he made his way to Casablanca and thence to England by ship. He survived the war and passed away aged eighty-nine.

(Courtesy of Mrs Edith Thomas)

Top: A British fighter pilot during 1940 showing resilience and determination. This is Flight Lieutenant Walter John Lawson of 19 Squadron who hailed from Somerset, but spent most of his early life in Kent. He joined the RAF in early 1929 starting as an aircraft apprentice and had become an aircraft fitter by 1931. He underwent training to become a pilot in 1936 and after achieving this was later commissioned in April 1940 and sent to 19 Squadron.

He fought over Dunkirk and was made a flight commander by September of that year before being awarded the Distinguished Flying Cross on 26th November 1940. By 27th June 1941, he had accounted for six enemy aircraft destroyed and one shared. He sadly failed to return from a sweep sortie escorting Blenheims on a low-level raid on shipping in Rotterdam harbour, when his aircraft was hit by enemy ground fire and he was killed aged twenty-eight. *(IWM CH 1362)*

Bottom: Pilots of 310 Czech Squadron pictured preparing for operations. August 1940. Left to right standing: Sgt Dvorak, Sgt Furst, Sgt Zima, P/O Fechtner, ? , Sgt Kaucky, Sgt Seda, ? P/O Goth. Standing over 3rd from left F/Lt Sinclair. Kneeling: Sgt Puda, P/O Boulton, F/Lt Jefferies. *(Author via P. Burgess)*

GB 1069 bc   Maßstab etwa 1:17000   1Km   Duxford
500m   Fliegerhorst

(1cm : 170 m)

G B
Gehe

Kriegs
0916

Nach
7.9.4

Karte
1:10
Blatt

Läng
(ostw.
0°08
Nördl
52°0

Zielh
über

Panzerabwehrgräben

Panzerabwehrgräben

Duxford

ca. 13 km nach Cambridge

G.B. 10 69   Fliegerhorst
1) 3 Flugzeughallen
2) 1 Flugzeughalle, zerstört
3) Unterkunftsgebäude, zerstört
4) Splittersichere Abstellplätze f.Flugzeuge
5) Blechstellungen

A German reconnaissance photograph showing a possible target and its surrounding area, dated 7th September 1940.
(IWM MH 26526)

Top: Czechoslovak pilot Sergeant Bohumir Furst returns from another successful patrol over the south-east. Furst destroyed three enemy aircraft between 3rd and 15th September 1940 and was later awarded the Czech Military Cross. He survived the war. *(IWM CH 1294)*

Bottom: Squadron Leader Jefferies (left) and Flight Lieutenant Sinclair read telegrams of congratulations to the Czechoslovak pilots of 310 (Czech) Squadron following their successful engagements against the enemy. The telegrams were sent from Archibald Sinclair, Secretary of State for Air, Dr Benes, Czech Minister and Air Chief Marshal Sir Hugh Dowding, Commander-in-Chief of Fighter Command. September 1940. Left to right back row: Sgt Bohumir Furst, Sgt Rudolf Zima, Sgt Raimund Puda. Front: F/Sgt Josef Komiinek, S/Ldr Jefferies, P/O Vaclan Bergman, F/Lt Sinclair, ? , Sgt Jan Kaucky. *(IWM CH. 1286)*

Top: Czech and English pilots give accounts of their operations against the enemy to the intelligence officer on return from combat. Left to right: Sgt Josef Rechka, F/Lt George Blackwood, P/O Jaroslav Sterbeck, P/O Emil Fechtner, F/Lt Gordon Sinclair and F/Sgt Miroslav Jiroudek. Intelligence officer's name unknown. *(IWM HU40538)*

Bottom: A view of one of 19 Squadron's Nissen huts with very basic amenities in 1940. *(Courtesy of Martin Sheldrick)*

Top: Seated and awaiting further information in the Duxford operations room, WAAF plotters are in constant communication with their controller and Observer Corps.

Bottom: A view of the interior of the station orderly room with RAF, WAAF and civilian clerks at work. *(IWM CH 1388)*

Top: One of the 'Few.' Pilot Officer Richard Leoline Jones in typical fighter pilot's flying kit, consisting of B-Type leather flying helmet, Mk II goggles, D-Type canvas oxygen mask, inflatable life preserver (mae-west) and 1936 pattern fur-lined leather flying boots. Jones joined 19 Squadron from 64 Squadron in September 1940 and flew in the Duxford Big Wing formation. He later rejoined 64 at Hornchurch in November that year.
(Richard Jones Collection)

Bottom: Spitfire Mk 1As of 616 Squadron about to take off on an early evening sortie from Fowlmere in September 1940. They are wearing the code letters QJ, which was at the time confusing as 92 Squardon also used the same code. *(IWM CH 1450)*

Top: Pilot Officer Frantisek Dolezal of 19 Squadron seated in his Spitfire. He joined 19 in early August 1940 and claimed a Bf109 on 5th September. He claimed another three destroyed and one damaged within the next two weeks. Dolezal was posted to 310 Squadron in 1941 and became its commander from April 1942 and awarded the DFC. He led the Czech Wing in 1943 but was tragically killed in a flying accident in Czechoslovakia in October 1945. *(Richard Jones Collection)*

Bottom: Bader is standing with two of 242 Squadrons aces. Left: Pilot Officer William Lidstone McKnight DFC and Flight Lieutenant George Eric Ball DFC. Both were killed later during the war. *(IWM)*

Top: A newly arrived Spitfire Mk IIA P7420 for 19 Squadron is refuelled from the tractor–towed fuel bowser at Fowlmere, Duxford's satellite airfield during September 1940. This aircraft was one of the few Mk IIs to see action with front–line squadrons before the end of the Battle of Britain. Note it still has to have the squadron identification code letters painted on to the fuselage. *(IWM CH 1357)*

Bottom: Pilot Officer Norman N. Campbell of 242 Squadron from Ontario, Canada pictured here in September 1940 seated on his Hurricane. He was lost over the North Sea off Yarmouth after engaging a Dornier 17 bomber on 17th October 1940. His body was recovered later and buried at Scottow Cemetery, Norfolk on 31st October. *(IWM CH 1409)*

Top: Pilot Officer Frantisek Hradil poses with his aircraft at Fowlmere. He joined 19 Squadron from 310 Czech Squadron on 28th August 1940. A ground-crewman is at work in front of the Spitfire. Hradil was killed in action during combat with German fighters over Canterbury on 5th November 1940, his Spitfire P7545 crashing into the sea off Southend. His body was washed ashore a week later and was buried at the Sutton Cemetery Southend. *(Richard Jones Collection)*

Bottom: Sergeant Pilot Alexander 'Mac' MacGregor of 19 Squadron wearing his leather Irvin jacket alongside his aircraft Spitfire P7547. He joined the squadron on 27th September 1940, remaining with it until 5th May 1941, when he was posted to 46 Squadron who set sail for Malta later that month. He survived the war and died in 1995. *(Author via Pat Burgess)*

Flying Officer Leonard Haines joined 19 Squadron in early 1940. Posing in his Spitfire in October 1940 after being awarded the Distinguished Flying Cross, Haines became an ace with eight and four shared enemy aircraft destroyed between 1st June and 28th November 1940. He was sent to instruct at 53 Operational Training Unit at Heston and was tragically killed in a flying accident on 30th April 1941. *(Author via P. Burgess)*

Top: Dr Eduard Benes awards medals to Czech airmen inside one of the hangars on his second visit to Duxford on 14th December 1940.
(IWM HU 40548)

Bottom: Spitfires of 19 Squadron pictured at Fowlmere during the winter of 1940.
(Author via P. Burgess)

Top: King George VI conferring a Bar to Flying Officer Albert Gerald Lewis in an awards ceremony at Duxford on 16th January 1941. A South African, Lewis had just returned to service with 249 Squadron, after being shot down and badly burnt on 28th September. By that time he had shot down eighteen aircraft. *(IWM CH1948)*

Bottom: 19 Squadron was still operating from Duxford's satellite at Fowlmere at the start of 1941 and continued to operate from the airfield until August that year before leaving for Matlaske. Pilot Officer Wallace Cunningham DFC is seen here with Flight Lieutenant Walter 'Farmer' Lawson outside one of the camouflage Nissen huts in early 1941.

(Courtesy of Tom Stevens)

Top: A group photograph of the commanding officer and his test pilots and technicians of the Air Gun Mounting Establishment (AGME). Back row 2nd left: Flying Officer Peter Pool. Front row 2nd right: Squadron Leader John Munro. Both these men fought in the Battle of Britain.
(Courtesy of A.P. Pool)

Bottom: A Grumman Martlet aircraft used by the Air Gun Mounting Establishment who were stationed at Duxford in early 1941. This American-built aircraft was more commonly known as the Wildcat, but in the Fleet Air Arm as the Martlet. *(IWM HU 41585)*

Opposite page: Flying Officer Peter Desmond Pool of AGME pictured at Duxford in 1941. Pool had joined the RAF in April 1937 and was commissioned in August 1940, going to 266 Squadron on 26th August before joining 72 Squadron at Biggin Hill on 3rd October. He was shot down in combat with enemy fighters on 11th October over Deal and was wounded, but managed to bale out of his Spitfire K9870 which crashed at Milton Regis, Sittingbourne, Kent. After recovering in hospital from his wounds, he was sent to fly aircraft for evaluation at the Air Fighting Development Establishment (AFDE) at Duxford in early 1941. He remained there until 12th August before being posted to 610 Squadron at West Malling. He was killed in action during the Dieppe Raid on 19th August 1942 while flying Spitfire EP235 coded DW-F.

(Courtesy of A. P. Pool)

Top: Flight Lieutenant Walter 'Farmer' Lawson takes control? Lawson is pictured with the local milkman's horse and dray which daily came from the Walston Dairy at Thriplow to Fowlmere in 1941. The brick barn behind was used by 19 Squadron as a mess during their stay and still stands today.

(Courtesy of Martin Sheldrick)

Bottom: F/Lt W.T. 'Farmer' Lawson DFC was later killed in action in 1941.

(Courtesy of Tom Stevens)

Top: Pilots of 56 'Punjab' Squadron 'mapping' their course prior to taking off. *(IWM CH 457)*

Bottom: 'The Sergeants'. Pilots of 19 Squadron caught in a completely informal pose. Left to right: Harry Charnock DFM, Kopriva, Stanislaw Plzak (killed in action 7th June 1941), Brooker, Brown, Kosina and Thomas. *(Courtesy of Tom Stevens)*

Opposite page: Pilot Officer Tom Stevens of 19 Squadron. Stevens was posted to the squadron from 602 Squadron at Prestwick on 24th December 1941. He flew on many offensive sweeps as the RAF brought the fight back across the Channel into France and Belgium against the Germans. He remained with the squadron until July 1941. *(Courtesy of Tom Stevens)*

Top: Pilots of 56 Squadron at Duxford in June 1941. Some of their Hawker Hurricane aircraft were donated by the Province of the Punjab. Their squadron commander, Squadron Leader Hugh Dundas DFC is holding the squadron mascot, Robin. *(IWM CH 4521)*

Below: Officers, airmen, WAAFs and civilian technicians of the Air Fighting Development Establishment and the Air Gun Mounting Establishment outside Hangar 3, which they both shared, October 1941. Those known: Alec Gray 7th from right top row. Second row standing 6th from left: Pilot Officer Ralph 'Titch' Havercroft. Front row seated 2nd right: Flying Officer Peter Pool, 6th from right: Squadron Leader John Munro commanding officer of AGME. *(Courtesy of A.P. Pool)*

Opposite page, top: A strange sight on the aerodrome? A Cierva C30A Autogiro of the Radar Calibration Flight 74 (Signals) Wing is wheeled out ready for take-off. Their role was so that defensive radar stations could precisely calibrate their equipment against their airborne positions. Although it looked very much like a helicopter, the aircraft could not hover. *(IWM CH 1426)*

Six Hawker Hurricane Mk IIBs of B Flight, 601 Squadron flying in starboard echelon formation over Thaxted, Essex.
(IWM CH 3517)

Top: Squadron pilots of 601 Squadron, the only RAF Squadron to be equipped with the American-designed Bell P39 Airacobra aircraft. Taken at Duxford in August 1941. *(IWM HU48131)*

Bottom: An Airacobra of 601 Squadron is helped by ground crew as it taxies in after a flight. *(IWM HU48127)*

Top: A Hawker Hurricane awaits testing with the AFDE in Hangar 3. *(Courtesy of A.P. Pool)*

Middle: A Hawker Typhoon pictured at Duxford in September 1941, where it would have been tested by the Air Fighting Development Establishment. The aircraft, either R7580 or R7581, was one of the first 15 Typhoon Mk 1As and had a three-bladed propeller. Later models were fitted with four blades. Note that the aircraft has not been fitted with armament. The standard weaponry for this period would have been an A-Wing consisting of twelve machine guns. *(Courtesy of Gladwin-Simms Collection)*

Bottom: An excellent aerial view of Hawker Typhoon F Mk 1Bs of 56 Squadron. Aircraft known are US-Y Serial No. R8825, US-C Serial No. DN317, US-H Serial No. R8824.

(Jacques Tremp)

Top: A captured Messerschmitt Bf109F aircraft arrived at Duxford along with other enemy aircraft types, underwent evaluation and was used for demonstration. The unit involved with this task was 1426 Flight. (Note the RAF markings on the wing and tail-fin).
(IWM HU 41591)

Bottom: Station commander Group Captain John Grandy (left) chats with squadron commanders, Squadron Leader Denys Gillam and Squadron Leader Bertie Wootton. Gillam commanded and led the first Hawker Typhoon Wing formed at Duxford in March 1942.
(IWM HU 42317)

95

Top: The RAF Duxford station band photographed in 1942. *(Courtesy of Old Duxford Association)*

Bottom: Pilots of the Air Fighting Development Establishment pictured in front of a De Havilland Mosquito fighter-bomber aircraft, which was undergoing test trials at Duxford during the early part of 1942. Flying Officer Peter Pool, 3rd from left back row. Squadron Leader John Munro front row centre. *(Courtesy of A.P. Pool)*

Spitfire Mk V AB505 (*top* and *above*) underwent trials with the AFDU in April 1942, after it had been converted to a HF IX with a Merlin 61 Rolls-Royce engine. This aircraft was being tested by Flying Officer Peter Pool. *(Courtesy of A.P. Pool)*

CHAPTER FOUR
THE AMERICAN YEARS
1943 – 1945

From the beginning of January until 26th February, the 350th Fighter Group continued at Duxford until it, along with sixty-one aircraft, was posted overseas to join the United States 12th Air Force operating in North Africa.

During February and March, other units also began to depart from the aerodrome to other airfields. The Air Fighting Development and the Naval Air Fighting Development Unit both moved to Wittering, 1426 Flight went to Collyweston and 1448 Flight to Heston, 169 Squadron being despatched to Barford St. John.

Plans were being detailed however that the aerodrome would be handed over to the 8th United States Army Air Force and become home to one of the fighter groups now preparing to escort the American bombers over northern France in daylight bombing raids. So it was that on 24th March 1943, Duxford aerodrome became home to the 78th Fighter Group, when an advance party arrived to start setting up headquarters and planning further construction of buildings. Between the 1st and 6th April the fighter group flew in seventy-five Republic P-47C Thunderbolt aircraft.

The Group consisted of three fighter squadrons, the 82nd, 83rd and 84th. The total of service personnel was 1,700. The station call-sign was Rutley. On 7th April, Marshal of the Royal Air Force, Lord Hugh Trenchard accompanied by General Hunter of the United States Army Air Corps visited the station and gave a talk to the American personnel.

The Group became operational within the month and undertook its first mission at noon on 13th April, when Lieutenant Colonel Arman Peterson led the 83rd Squadron in a sweep over north-east France. The trip was uneventful apart from one pilot, Lieutenant Colonel Joseph Dickman having to bale out into the sea, on the return leg of the mission, fortunately being picked up by RAF air-sea rescue craft.

On 14th May, the mission of the day was to act as escort to bombers whose targets were at Antwerp. While over the Antwerp area, the squadrons were engaged by more than twenty Focke-Wulf and Messerschmitt 109s. It was during this action that the Group scored its first enemy victories, when Captain Robert E. Adamina of 82nd and Major James J. Stone commanding officer of 83rd, each claimed a

Focke-Wulf destroyed. Three pilots of the 78th were shot down during the encounter including Captain Adamina, but none were killed.

The aerodrome played host to royalty on 26th May, when King George and Queen Elizabeth arrived to visit and meet the Americans of the 78th. The King and Queen were received by senior American officers, met their aircrews and inspected their aircraft.

On 15th June Duxford was officially handed over by the Royal Air Force to the 78th Fighter Group as Station 357 of United States Army Air Force 8th Fighter Command, when in an official ceremony Wing Commander Matthews presented Colonel Peterson with a silver vase.

1st July was a bad day for the 78th Fighter Group, when it lost its highly respected leader Lieutenant Colonel Peterson in action, although its pilots destroyed four Focke-Wulfs and damaged five others that day. Peterson's replacement, Lieutenant Colonel Melvin McNickle arrived to take command on 12th July.

For the next month, the 78th continued its role as fighter escort or cover for the bomber raids over Europe and was beginning to earn itself an excellent reputation for some successful missions. Squadron morale was high during this period and many celebrities of the time visited the aerodrome, including Bob Hope, Francis Langford, James Cagney and Bing Crosby. Radio broadcasts were also recorded as was the 78th's own dance band, the 'Duxford Thunderbolts' and some transmitted to audiences in the United States.

At the end of July, the Thunderbolt aircraft were fitted for the first time with fuel drop tanks, which meant that they could now escort the bomber formations further into Germany.

The Group provided return cover for bombers on a raid to Kassel on 30th July and during the sortie claimed sixteen victories, becoming the first American unit to claim double figures in one mission. Other firsts that day included Captain Charles P. London of 83rd Squadron who claimed two enemy aircraft destroyed which made him the first American ace in the European war with five victories. One pilot Major Eugene Roberts became the first American to claim three enemy aircraft in one mission and was awarded the American Distinguished Flying Cross. Lieutenant Quince Brown was recorded as the first American Air Force pilot to ground-strafe a locomotive and gun-battery after he had suffered engine problems and had to fly home at very low altitude. Unfortunately, the new commanding officer Colonel McNickle was shot down that day, but survived to become a prisoner of war. His position was taken by Lieutenant Colonel James J. Stone.

The aerodrome was visited by the commander-in-chief of the United States Army Air Force, General Hap Arnold on 4th September 1943. He was shown around various buildings and watched a display from the control tower of the latest British and US fighters. General Arnold later during his visit gave a lecture on the history and future role of the American Air Force.

For the remaining part of the year the squadrons continued with the tiring role of escort duty, many sorties lasting over five hours in the air, which was extremely exhausting for the young pilots, who had to have full concentration. Any minute during the approach or return leg of the mission, German fighters could appear and attack.

The 78th's tactical role was to change at the

beginning of 1944, when it was given new orders to fly fighter-bombing missions. The task was to carry two small bombs, one under each wing and after dropping the load on the target to continue hitting the targets with gun strafing and low-level attacks or by engaging in air combat.

The Group carried out its first fighter-bomber mission on 25th January 1944, to attack a German airfield in occupied France, but the mission failed due to bad weather. On 31st January, the mission was far more successful however, when the Group dropped thirty-five 500-pound bombs onto the airfield runway at Gilze-Rijen in Holland causing extensive damage.

During the latter part of February, the missions reverted to escort duty, when operations were detailed to attack the industrial ball-bearing factories. On 6th March 1944, the Americans launched their first large-scale raid on Berlin, in total 702 heavy bombers and an escort force of 644 fighters from both 8th and 9th Air Force. Duxford would provide escort for the 1st Bomb Division's B17 Flying Fortress aircraft whose primary target was the VKF ball bearing works at Erkner. Entering and returning from the target area the German Luftwaffe and ground defences put up a tremendous fight that day and American casualties were enormous, losing sixty-nine heavy bombers and eleven fighters. The 78th lost two pilots, Flying Officer E. Downey, who was killed and 2nd Lieutenant G. Turley. The raid was later recorded as the largest loss of aircraft and aircrews suffered throughout the entire air campaign by the 8th Air Force.

A few miles away, on 5th April, Fowlmere became the home of the 339th Fighter Group, part of 66th Fighter Wing of 3rd Bomb Division. The 339th consisted of the 503rd, 504th and 505th Fighter Squadrons and the aerodrome was designated as Station 378 with call-sign 'Gas Pump.'

The Group operated North American P51 Mustang fighter aircraft. New buildings had been erected on the site including a T2 hangar, seven Blister hangars and Nissen huts. Two PSP runways were also laid, one running from north-east to south-west at a length of 1,600 yards, the other on an east to west axis at 1,400 yards.

During a raid on 19th May, when escorting B-24 Liberators, the Duxford squadrons engaged enemy fighters over Brunswick, Germany and after the combat claimed ten Messerschmitt 109s and two Focke-Wulfs destroyed for no loss. These claims brought the 78th's victories past the 200 mark. On 22nd May the Group's commanding officer Colonel Stone handed over his position to Lieutenant Colonel Frederick Gray.

During this month, Duxford along with other aerodromes, especially in the south-east of England, started to increase security as preparations began for the build-up to the forthcoming Allied invasion of Europe. Several wing commanders of 66th Fighter Wing flew into Duxford for planning and briefings at Sawston Hall Headquarters which was located nearby.

The invasion was scheduled for 5th June and Duxford squadrons would take part in bomber support and ground attacks against German coastal defences along the French coast in the days leading up to the assault.

The Newmarket to Royston road was sealed off between the aerodrome and the domestic site with guards and special wardens on duty and all civilians were kept out from the surrounding area. As D-Day approached all the aircraft were painted with a special recognition black and white stripe over the top and underside of the wings as well as on the fuselage.

The operation codenamed Overlord was cancelled for twenty-four hours due to bad weather conditions in the Channel, but was then given the go-ahead by supreme commander of the Allied forces, General Dwight Eisenhower for 6th June 1944.

Early that morning at 3.30 am Thunderbolts from Duxford took off on their first mission of the day to provide cover for the landing craft now nearing the Normandy beaches. During their first sweep over the beachhead they encountered no enemy aircraft and returned to Duxford to refuel and re-arm to bomb any German targets of opportunity. The 84th Squadron attacked enemy targets in the Alençon area including an ammunition dump, two locomotives with box cars and flat wagons. The 82nd, 83rd and 84th all carried out missions that day over the invasion area, the final sortie landing back at Duxford at 11.00 pm. As the Allied ground troops slowly moved in land, the squadrons continued to support the advance.

On 9th June the 78th became unexpectedly heavily engaged with enemy fighters whilst it was dive-bombing German positions. The combat was fierce and the Group lost ten pilots during the engagement including the commanding officer of 84th, Major Harold E. Stump. These were the highest casualties that the 78th had suffered since arriving at Duxford two years previously. By the end of that month the Group had flown forty-five missions over the Normandy area. The month of July continued with operations moving further into

France as the Germans retreated mile by mile.

Tragedy was to strike Duxford on the evening of 19th July 1944 however, when a visiting B17 Flying Fortress of 612th Bomber Squadron, 95th Bomber Group flew over the aerodrome. The pilot of the aircraft, 1st Lieutenant James D. Sasser had flown in to visit friends who served with the 84th Squadron. During the aircraft's approach, the pilot buzzed the control tower in a display of low-flying and while pulling the aircraft back up over one of the hangars, hit one of the obstruction lights on the hangar itself. The aircraft's left wing and horizontal stabilizer were torn off and the bomber rolled over and crashed to the ground, just missing the officers' barrack block, but hitting the main barracks of the 83rd Fighter Squadron and part of the 82nd's. All twelve on board were killed in the explosion and one airmen of the 84th, 1st Lieutenant Martin H. Smith.

Quickly on the scene of the crash was the chaplain, Captain William J. Zink, who tried to rescue a man trapped in the barracks, but after two unsuccessful attempts, his way barred by falling beams and heat from the fire, he was forced to give up. For his actions that day, Zink was awarded the Soldiers Medal, becoming the first 8th Air Force chaplain to receive such an award. It was put on record that if the crash had occurred thirty minutes later, the death toll would have been much higher, as many servicemen would then have finished duty.

A new record was set by Duxford's 82nd Fighter Squadron on 28th August when it claimed the first destruction of one of Germany's new revolutionary jet fighter aircraft, the Messerschmitt Me262. The squadron had been flying west of Brussels, Belgium, when two pilots Major Joe Myers and

Smoke and flames cloud the sky after a B17 Flying Fortress crashed into one of the barrack buildings on 19th July 1944, after accidentally hitting an obstruction light, whilst buzzing the aerodrome. All aboard the aircraft were killed with one fatality on the ground. *(Copyright USAAF)*

Lieutenant Manfred O'Croy sighted the jet aircraft flying fast and low beneath them. They began to give chase and after applying various aerial tactical manoeuvres forced the pilot to crash land his aircraft. The remainder of the squadron in turn destroyed the German aircraft by ground-strafing.

By September 1944, the Allied commanders were pushing the Germans to the borders of the Reich. One plan devised by British commander Field Marshal Bernard Montgomery was to quicken the end of the war by capturing a series of bridges in Holland; this would open up a back door into Germany's industrial Ruhr. The plan codenamed Market Garden would entail flying 35,000 Allied troops behind enemy lines to be dropped by parachute or land by glider, capture the target and then hold until XXX Armoured Corps pushed through the German

lines and relieved the airborne troops.

The operation began on 17th September and Duxford's squadrons were heavily involved with the support of the landings at the Nijmegen and Arnhem bridges. Their role was to divert enemy fire from the glider aircraft, and they carried out the operation so successfully that the Group was awarded a unit citation and commanding officer Colonel Gray the Silver Cross for his outstanding leadership. However, as is now well known, the attempt to capture the bridge at Arnhem ended in failure.

On the 23rd, the 78th Fighter Group were involved in action against enemy ground defences, and saved many Dakota C-47 and Short Stirling aircraft who were dropping much needed supplies over Arnhem. The American fighters met heavy barrage from German flak batteries, but managed to destroy enemy positions hidden in hedgerows and one concealed gun in a church tower. Despite their actions, Allied troops were evacuated on 26th September. The gamble to end the war early had failed.

Duxford was to undergo further changes during the latter part of the year. During wet weather the aerodrome's grass flight-paths were always prone to flooding and muddy conditions causing unseen problems for aircraft, so much so that the 78th nicknamed it the 'Duckpond.' It was therefore decided to construct a Pierced Steel Planking Runway for this purpose. The PSP runway was to be 3,500 feet in length and 150 feet in width.

On 6th November 1944 at 3.20 pm a mid-air collision occurred near the aerodrome, when Captain George T. Zeigler of 2nd SF (Scouting Force) based at Steeple Morden, flying one of the newly arrived P51 Mustang aircraft D44-14898 which were being brought into replace the 78th Fighter Group's P47 Thunderbolts, collided with another Mustang. Zeigler, accompanied by Captain Counselman flying on his right, was on a training flight with four other P51s of 339th Fighter Group, but as they crossed Duxford at 8,000 feet, a P47 Thunderbolt appeared unexpectedly causing Zeigler to break right and collide with one of the other Mustangs. The starboard wing of Zeigler's aircraft was torn away and he went into a slow spin to the right. He was forced to bale out and received minor injuries.

Unfortunately the pilot of the other Mustang, Lieutenant Alan Francis Crump from Michigan of 503rd Squadron was less fortunate and he was killed when his aircraft 44-14143 crashed near Manor Farm, Fowlmere. Crump was laid to rest at the American military cemetery at Madingley, Cambridge. Plot D, Row 6 Grave-25.

During the latter part of November the new runway construction was laid down, and the 78th moved out in early December to operate from Bassingbourn, ten miles to the west and the home of the 91st Bomb Group. By 11th December the work on the runway had been completed and the unit's aircraft returned.

On the last day of 1944, the 78th flew its final mission on P47s and claimed its 400th enemy aircraft destroyed, when Captain Julius P. Maxwell downed a Focke-Wulf 190.

With the arrival of another year, Duxford squadrons were about to be fully re-equipped with the North American P51 Mustang. Although the Mustang was the most up to date and advanced performance fighter, many of the pilots at Duxford wanted to continue to fly their trusty P47s. A few Mustangs had been flown in to the aerodrome for the pilots to convert to, but by 5th January 1945, all three squadrons at Duxford had been re-equipped.

As the Allies advanced into Hitler's crumbling Third Reich, the casualty rates increased. At Duxford, February was to prove the costliest period for the whole of the 78th's stay at the aerodrome with the loss of eighteen pilots missing in sixteen missions. In return the squadrons had destroyed 105 locomotives, nineteen oil tankers and many other vehicles. A change of command took place during the same month when Colonel Gray was posted to Headquarters 8th Air Force, being replaced by Lieutenant Colonel Olin E. Gilbert, but his position was soon taken by Lieutenant John D. Landers and Gilbert was made deputy commander.

By now most of the German Luftwaffe were grounded due to the lack of fuel and pilots, except for the odd occasion. On 10th April the

VICTORY!

May 8th 1945, Victory in Europe. American personnel at Duxford start their celebrations, which continued well into the next day. *(Copyright USAAF)*

Duxford, the airmen of the 78th celebrated all through the day and into the early hours.

During the two years that the 78th Fighter Group had flown operationally from Duxford, it had completed 450 operations, which amassed over 80,000 flying hours. During the missions they had destroyed 338.5 aircraft in the air, and a further 338.5 on the ground, totalling 697. Their own losses totalled 93 pilots killed and 167 aircraft.

Group destroyed fifty-two German aircraft on the ground and damaged a further forty-three, but the biggest tally was to be recorded when providing freelance support for a bombing mission attacking enemy airfields in eastern Germany and Czechoslovakia on 16th April 1945, when they claimed 135 aircraft destroyed and eighty-nine damaged. This was a record for strafing by any group of the 8th Air Force. With the defeat of Nazi Germany now at hand, the 78th's final mission of the war flying from Duxford was in co-operation with its British counterparts.

On 25th April, the 78th was one of two American 8th Air Force groups to provide escort to Avro Lancasters of 617 'Dambuster' Squadron on a raid to bomb Hitler's mountain home at Berchtesgarden. Weather conditions were bad that day with low cloud and the target could not be seen visually and so only the nearby SS Barracks was bombed.

Victory over Germany was proclaimed in Europe on 8th May 1945. Adolf Hitler had committed suicide in his bunker in Berlin and the world rejoiced at the fall of Nazism. At

With the war won, all that the American servicemen wished for was to return home to loved ones, but this would take time and it would be six months before the 78th would be finally ready to return personnel back to the United States.

The aerodrome held a special open day for the local public on 1st August. Approximately 5,000 people attended and met many of the servicemen who provided them with numerous gifts, the type of which the local population who had been restricted by rationing, had not seen for the past five years.

In October 1945, the 339th Fighter Group left Duxford's sister airfield at Fowlmere and soon arrangements were made to dismantle the site. By November 1945, most of the American personnel had left the aerodrome and only a small maintenance party remained. So ended another page in the history of Duxford aerodrome, when, on 1st December, it was officially handed back to the Royal Air Force.

Top: King George VI and Queen Elizabeth arrive at Duxford on 26th May 1943, to inspect the newly arrived United States Army Air Force personnel. Here they are greeted by Brigadier General Frank Hunter of the Eighth Air Force Command, outside the station headquarters. A guard of honour by the RAF Regiment stands in the background. *(IWM CH 19214)*

Bottom: Brigadier General Hunter shows the King and Queen one of the American P47 Thunderbolt fighter aircraft presently undergoing maintenance. *(IWM HU41594)*

Top: Wing Commander S.L. Matthews, station commander at Duxford, presents a silver presentation bowl to Colonel A. Peterson of 78th Fighter Group at the official handover of the station to the United States Army Air Force in June 1943. *(IWM HU 51427)*

Bottom: Pilots of 84th Fighter Squadron in front of one of their Republic P47 Thunderbolt aircraft, June 1943. *(IWM)*

Top: Armourers at work on two of the eight .50 caliber Browning machine guns which were housed within the P47's wings. *(IWM HU 48143)*

Bottom: A complete engine change of the massive Pratt and Whitney Twin Wasp 2800 engine is being undertaken on this P47. *(IWM HU 48138)*

Colonel Arman Peterson, commanding officer of 78th Fighter Group. He led the group from April 1943 until he was killed in action on 1st July 1943. *(Author via J. Cook)*

Top: P47s prepare to take off on another mission. The aircraft in the foreground was flown by Captain Jack Price. *(Author via J. Cook)*

Bottom: The 78th Fighter Group's own dance band, 'The Thunderbolts' played numerous shows and parties at Duxford and also recorded sessions for transatlantic broadcasts. *(Copyright USAAF)*

Top left: Colonel James J. Stone who took over command of the 78th Fighter Group from 31st July 1943, and would lead it until 22nd May 1944. *(Copyright USAAF)*

Top right: Captain Jack Clayton Price of Grand Junction, Colorado in the cockpit of his P47 'Feather Merchant II'. He served as commanding officer of 84th Fighter Squadron from 28th September 1943 until 25th February 1944, surviving the war and remaining in the air force until retiring in 1968. Price scored a total of five air victories. *(Author via J. Cook)*

Bottom: Lieutenant William S. Swanson in the cockpit of his P47, in January 1944 after destroying a Focke Wulf 190, his first victory. *(IWM EA 13448)*

Top: Two aces of 78th Fighter Group. Major Eugene P. Roberts of 84th Fighter Squadron is seated in his personal P47 'Spokane Chief.' Beside him is Captain Charles P. London of 83rd Fighter Squadron. Roberts achieved nine victories and one probable while London claimed five destroyed, two damaged and one probable. *(IWM HU 31360)*

Bottom: An impressive three squadrons of 78th Fighter Group's P47 Thunderbolt aircraft lined up in 1943. *(Author via J. Cook)*

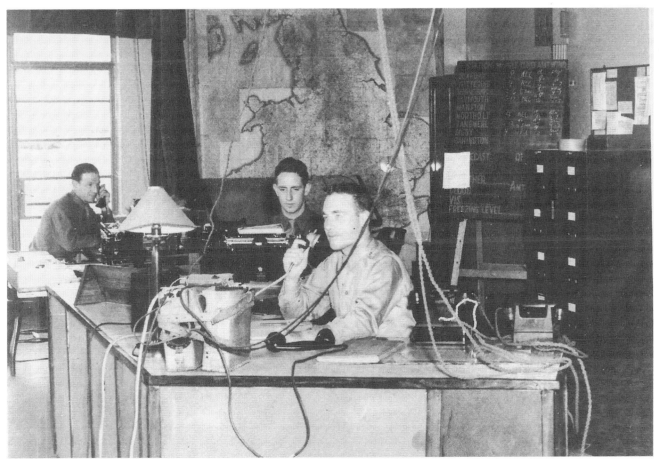

Opposite page: A tremendous photograph of Captain Charles P. London in his P47. *(Author via J. Cook)*

Top: The American flag flying at the main gate entrance to Duxford, circa 1944. *(IWM HU 40429)*

Bottom: One of the communication rooms manned by American personnel. *(IWM HU 41205)*

Top: Lieutenant General Ira C. Eaker, Commander-in-Chief of 8th Air Force (centre in Macintosh coat) visits the aerodrome with other senior officers and inspects a P51 Mustang, 1944. *(IWM HU 31913)*

Bottom: A view from the control tower. Lieutenant General Eaker (second right) with senior American Air Force officers. *(IWM HU 31908)*

Top: Officers of 82nd Fighter Squadron, Independence Day, 4th July 1944. Left to right on wings: L. Dicks, L. Hosford, H. Becks, H. Slater, M. Capp, G. Lundigan, W. Young, W. Vandyk, H. Morris, J. Kinsolving. Standing left to right: W. Guilfoyle, R. Miller, C. DeWitt, B. Watkins, R. Boeckman, B. Smith, Colonel Frederick C. Gray (commanding officer, 78th FG), Ben Mayo (commanding officer, 82nd FS), M. Woller, R. Wilkinson, R. Bosworth, R. Wolfe, W. Brown, J. Fitzgerald, H. Shope and R. Bosworth. *(IWM HU 65339)*

Bottom: Pilots of 82nd Fighter Squadron relax outside the 'ready room' during the summer of 1944. Left to right: Eggleston, R. Wolfe, W. Brown and Nelson. *(IWM HU 31937)*

Top: This two-seater P51 Mustang 'Gruesome Twosome' was for use by senior officers of USAAF in 1944. *(IWM HU 31936)*

Bottom: The American film star James Cagney chats with an airman during his visit to Duxford in 1944. *(IWM HU 31915)*

Top: Comedian and film star Bob Hope and actress Francis Langford seen at Duxford in front of P47 'Vee Gaile', the aircraft used by Major Robert E. Eby.

(Bob Bryant Collection)

Bottom: The control tower, signal square and hangars in 1944.

(IWM HU 31938)

117

Top: A good landing is one you can walk away from? Lieutenant Peter W. Klaassen of Detroit poses for the camera next to P47 'Noamie Vee' following a belly-landing at Duxford on 5th September 1944. The Thunderbolt's pilot was Lieutenant Franklin R. Maywood of 83rd Squadron. The aircraft was repaired and sent to 23rd Squadron of 36th Fighter Group, but it was lost in action on 19th March 1945.

(Author via J. Cook)

Bottom: Major Richard A. Hewitt in front of his P47. From Lewiston, New York, he served with the 82nd Fighter Squadron.

(Courtesy of Peter Randall via Richard Hewitt)

Top: P51D Mustang 'Mountaineer' of 357th Fighter Group after crash-landing during the winter of 1944. Note the matting used on the airfield at this time. *(Author via J. Cook)*

Bottom: American airmen enjoy the winter snow, late 1944. *(IWM HU 31932)*

Top: The P51D Mustang 'Big Beautiful Doll' flown by Lieutenant Colonel John D. Landers, who commanded the 78th Fighter Group from 22nd February to 28th June 1945. During his time at Duxford, he recorded three Messserschmitt 109s destroyed and a shared destroyed on a Me262 German jet fighter.
(IWM HU 48197)

Bottom: 'Heavenly Body'. Lieutenant L. Nelson of 82nd Fighter Squadron in his P51 Mustang.
(IWM HU 31924)

Bottom left: Lieutenant Huie H. Lamb Junior of Abilene, Texas in his Mustang 'Etta Jeanne II', P51K 44-11631. He named the aircraft after his younger sister. *(Courtesy of Peter Randall via Lt H. Lamb)*

Bottom right: Wing Commander Alan Deere, a New Zealand fighter ace during WW2, had joined the RAF in 1938. He became Duxford's first postwar station commander, on 1st December 1945. *(Author)*

Top left: Lieutenant Antony P. Palopoli who hailed from Pittsburgh PA with his ground crew and P51D 'Little Joe'. Palopoli served with the 83rd Fighter Squadron. *(Courtesy of Peter Randall via Cathy Palopoli)*

Top right: P51 Mustang 'Lady Eve' with her pilot Lieutenant Burton J. Newmark of Brooklyn, New York. He flew with the 84th Fighter Squadron. *(Courtesy of Peter Randall via B. Newmark)*

Chapter Five
THE JET ERA
1946 – 1962

With Duxford now back with the RAF, the first preparations were put into action to receive operational squadrons. The first postwar station commander was Wing Commander Alan Deere – an extremely experienced leader and well liked and respected amongst his officers and men. The first RAF squadron to return to Duxford was 165 with Spitfire LFIXs and by April a further squadron 91 also flying Spitfires had arrived. On 8th June 1946 a massive victory flypast took to the skies and headed for London. The formation of aircraft included the two Duxford squadrons.

91 Squadron briefly left Duxford for Lübeck, Germany on training exercises in August, but had returned by 1st September. 165 Squadron had in the meantime disbanded and then re-formed as 66 Squadron, and it too headed for Germany.

Although the Royal Air Force's main aircraft were still piston-engine, the newly developed jet engine fighters were now becoming more available and soon the squadrons were being converted to such aircraft.

Duxford took delivery of its first jet aircraft on 30th October 1946, when a Meteor F3 arrived on charge to 91 Squadron. On returning from Germany, 66 Squadron also began to convert to the Meteor aircraft.

On 31st January 1947, 91 Squadron was renumbered 92. (92 had previously been one of the top squadrons throughout the war, especially over Dunkirk and during the Battle of Britain.)

During the next few months the aerodrome saw other squadrons come and go. Some of these squadrons had seen previous service at Duxford during World War Two, including 56 Squadron which arrived on 17th April. In September 74 and 222 flying Meteors landed to stay briefly for the Battle of Britain Flypast as did two piston-engine aircraft squadrons, numbers 3 and 80, flying the Hawker Tempest.

Problems with Duxford's runway were again under examination at the beginning of 1948 as the pierced steel planking was a cause of anxiety for the jet aircraft. The Air Ministry brought in contractors to update the runway and while this work was undertaken, the squadrons moved out temporarily to Martlesham Heath. No sooner had the work been completed, when under Air Ministry orders, the plans were totally revised with a concrete runway to be built running alongside the steel planking strip.

During June, Duxford took part in tactical exercises in order to gauge how the United Kingdom's air defence system would stand up to an enemy attack. The results were not good and the Air Ministry ordered a review.

Duxford aerodrome would have a new runway constructed and extra land would be purchased in order to build towards the eastern end of the aerodrome. Duxford squadrons 66 and 92 were moved to Linton-on-Ouse in Yorkshire, and the aerodrome was placed on 'care and maintenance'.

Construction of the new runway was finally started after months of delay on 18th September 1950 by the contractors W & C. French to build a concrete runway of 6,000 feet in length and a new perimeter track with hardstandings at various points. The plans also included the building of a T2 hangar at the eastern end to allow more maintenance area and an operational readiness platform which widened each end of the runway. The new runway was finally declared operational in August 1951.

Two new squadrons were posted to Duxford, 64 and 65 both using Meteor F8s. Soon the squadrons were undergoing exercises and scrambles for the new 'Quick Reaction Alert System' that had been devised in response to the possible threat from Russian air strikes as the Cold War materialised between the countries of western Europe, the United States and Communist Russia.

Tragedy was to strike 64 Squadron on 29th February 1952, when three of the squadron's aircraft were lost. As bad weather drew in the Meteor 8s were recalled by Duxford and told to divert to the nearest airfields. One Meteor which suffered a damaged undercarriage on take-off had to force land at Wattisham in Suffolk. The pilot Flying Officer F.B. Monge survived.

The pilot of the second aircraft Sergeant Verrico was ordered to bale out in 8/8ths cloud over Saffron Walden in Essex. Sadly the third pilot Flying Officer J. Catchpole became disoriented and went into a vertical dive from which he never recovered. His Meteor WE937 crashed near Debden airfield, Essex. Catchpole was killed instantly and he was buried at Whittlesford cemetery.

The aerodrome was involved in another large formation flypast on July 1953 for Queen Elizabeth II's Coronation Flypast over London. Two further squadrons, 3 and 67, flew in for temporary attachment with their Sabre jets (known as the F-86 in the United States Air Forces).

The Station Flight received a Spitfire LF 16 Serial No. TE357 in 1954 and it was used for flypasts and ceremonial duties. It was recorded in the station diary as 'the last Spitfire in Fighter Command' although this was not so. In August 1956, 64 Squadron was re-equipped with Meteor NF 12s and changed its role to that of a night-fighter squadron.

On 24th January 1957, the Station Flight's Spitfire was involved in an accident and was badly damaged and struck off charge, but a replacement aircraft was soon found, Spitfire SL542. In March that year 65 Squadron exchanged its Meteors for the new RAF fighter the Hawker Hunter F6.

On 14th September, the aerodrome held its annual Battle of Britain Day show. One of the highlights of the day's events was the display by the RAF's latest prototype aircraft, the English Electric Lighting. This aircraft could reach a speed of Mach 1 breaking through the sound barrier.

In 1958, 64 Squadron had replaced its old worn Meteors with Gloster Javelin FAW 7s (all weather aircraft). The routine of home and NATO exercises continued throughout that year.

In the July of 1959 three officers from the station were part of the four-man entry by the Royal Air Force into the *Daily Mail* newspaper's Blériot Anniversary Air Race between Paris and London. The race consisted of various transport modes including running, motor cycles and of course aircraft. Duxford was highly delighted when the winner of the race was Squadron Leader C.G. Maughan of 65 Squadron who completed the journey with a winning time of 40 minutes and 44 seconds.

On 6th July 1960, a special double ceremony was held for both 64 and 65 Squadrons, when they were presented by Marshal of the Royal Air Force Sir William Dickson with squadron standards after completing twenty-five years of service.

In 1960, the Air Ministry reviewed many of its airfields and many were marked for closure. Duxford was one of those being studied at this time. Plans were being looked at to extend the runway, but the cost of this and refurbishment to existing buildings deterred the Air Ministry and sealed Duxford's fate. Consequently January 1961 saw the beginning of the rundown of the aerodrome. 65 Squadron was disbanded on 31st March and the Station Flight in May. 64 Squadron was posted to Waterbeach on 28th July.

The final curtain fell on Duxford's flying history on 31st July 1961, when the last official flight was undertaken by Air Vice-Marshal R.N. Bateson the Air Officer Commanding 12 Group who took off in a Meteor NF 14. Thus forty-three years of flying service was brought to an end.

On 1st October the aerodrome was transferred to Flying Training Command, but this was to be brief. During this period the aerodrome was used by the RAF Gliding School with Chief Flying Instructor Max Bacon in charge. The aircraft used was the T21 Tutor and T31. The gliding school remained at Duxford until 27th April 1963.

The site was now left under the control of a care and maintenance party while decisions were pondered over what to do with Duxford by the Ministry of Defence. By 1962, the aerodrome began to show the first signs of neglect as weeds and grass began to take hold on the concrete apron and runway. It seemed that the aerodrome was to meet the same inevitable fate that had befallen other RAF bases during this period of cutbacks.

Right: Wing Commander Herbert Moreton Pinfold was station commander of Duxford from 4th May 1948 until May 1951. He had served with 56 Squadron as squadron leader during the Battle of Britain before going to 10 Fighter Training School Tern Hill as an instructor. He was at the RAF Staff College in 1945 and was later sent to the Far East on the staff of Headquarters Kandy, Ceylon and Singapore. Pinfold returned to command Duxford for a second tour in April 1956 till 1958, when he retired as group captain.

(Group Captain H.M. Pinfold Collection)

Top: Gloster Meteor jet aircraft of 66 Squadron lined up on the main apron.
(G/Capt H. M. Pinfold)

Bottom: Ground crew personnel working on a Meteor jet. Note the chap with the aviation fuel hose filling up the tank from the petrol bowser.
(IWM HU 41600)

Top: Pilots at the aerodrome have time enough before their next flight to take in a game of bowls, just in front of one of the Meteor jets.
(IWM HU 41599)

Bottom: Ground personnel of 66 Squadron relaxing while the squadron is up on exercise. 1949.
(G/Capt H.M. Pinfold)

A grand aerial shot of four Gloster Meteors which were part of a detachment of eight aircraft from 66 and 92 Squadrons stationed at Duxford, sent on a goodwill visit to Norway and Sweden in August 1949. *(G/Capt H.M. Pinfold)*

A frontal shot of the detachment from Duxford to Norway just after they left the ground. Two Dakotas and an Avro York transport aircraft with Squadron Leader G. Pinney in charge also took off carrying the ground servicing crews along with seven tons of spares. *(G/Capt H.M. Pinfold)*

Top: An Avro York transport aircraft arrives at the aerodrome, 1949. *(IWM HU 50660)*

Bottom: Men and women personnel of the RAF stores section at Duxford 1949. Left to right standing: S. Langham, B. Sharpe, L. Murray, J. Zebedee, N. Townsend, A. Scragg, B Sach. Front kneeling V. Weatherburn and Cpl F. Knowles. *(IWM HU 50654)*

Top: Members of the RAF Shooting Team which won the Nobel Cup 1952. Centre is Squadron Leader Bateson DSO, DFC.
(IWM HU 40613)

Bottom: Station commander Jamie Rankin DSO, DFC, checks his daily paperwork with his secretary. Rankin, an ex fighter pilot who fought during the Battle of Britain, commanded Duxford from 12th January 1953 until December 1954.
(IWM MH 28315)

An aircraftswoman pictured using equipment to check various aircraft clocks and dials in the instrumentation workshop.
(IWM HU 41603)

Top: The aerodrome played host to various VIPs during 1953/54, which included a visit by Emperor Haile Selassie of Ethiopia seen here inspecting aircraft of 65 Squadron. *(IWM HU41612)*

Bottom: Marshal Tito of Yugoslavia meeting with squadron officers. *(IWM HU 41615)*

Top: A view from inside the control tower as air controllers give their instructions. *(IWM HU 41611)*

Bottom: RAF personnel watch a fencing demonstration by fellow colleagues. Note the De Havilland Chipmunk aircraft behind. *(IWM MH 28305)*

Top: Duxford's Station Flight Spitfire SL542 pictured in 1956. *(Courtesy of Old Duxford Association)*

Bottom: Gloster Meteor F8 taxies past the station flight tower and Spitfire SL542. *(Courtesy of Old Duxford Association)*

Top: 64 Squadron pilots and ground crew pictured in front of their Gloster Javelin FAW Mk 7 aircraft in 1956. *(Courtesy of Old Duxford Association)*

Bottom: De Havilland Chipmunk aircraft WZ870 of the Station Flight. Jim Garlinge is seated in the cockpit with Robert Hope standing. *(Courtesy of Old Duxford Association)*

Top: A Hawker Hunter of 65 Squadron is prepared for operation, with a trolley carrying oxygen being pushed into position. Ground-crewman John Sills is standing on the cockpit ladder. *(Courtesy of Old Duxford Association)*

Bottom: A Gloster Javelin FAW7 of 64 Squadron taxiing into position to take off, 1958. *(Courtesy of Old Duxford Association)*

137

Top: Engineer Jim Garlinge at work on a Rolls-Royce Avon jet engine used on Hawker Hunter aircraft, 1958. *(Courtesy of Old Duxford Association)*

Bottom left: Station commander Group Captain Norman Ryder DFC is seated and hoisted aloft by fellow officers during evening jollities in the officers' mess in 1958. *(Courtesy of Old Duxford Association)*

Bottom right: Two Hawker Hunter aircraft of 65 Squadron in formation with an Avro Javelin of 64 Squadron fly over the aerodrome during a summer's day in 1959. *(IWM HU 46501)*

Five views of the aerodrome prior to its closure in 1961

Top: The main entrance. *(IWM HU 3865)*   *Bottom:* A good view showing the officers' mess with its well kept lawn and flower beds. *(IWM HU38654)*

Top: The final inspection parade of Royal Air Force personnel at RAF Duxford. *(IWM HU 38668)*

Left: The layout of the aerodrome's accommodation and office buildings, with the A505 top right. *(IWM MH 38663)*

Above: Across the concrete apron showing hangars and control tower. *(IWM MH 38658)*

Inset: The tower as it is today. *(Author)*

CHAPTER SIX
SAVED FOR THE NATION
1967 – PRESENT

In 1967, the Ministry of Defence was contacted by the United Artists Film Company seeking permission to use the aerodrome as one of the locations for a film about the Battle of Britain. After mediation with the MOD, and the owner of Manor Farm, Thriplow, C.R. Smith (see letter on page 145), the production company moved into Duxford during the spring of 1968, to prepare for the various filming sequences that would be shot at the airfield.

The aerodrome was recreated back in time to 1940 at a cost of £38,000 with camouflaged hangars and Spitfires and Hurricane aircraft and extras in wartime period costumes. The opening sequences to the film show a Hawker Hurricane squadron with a French château in the background, being strafed by low-flying Messerschmitts. This was filmed totally at Duxford, as was one of the airfield bombing scenes. The production company actually gained permission to destroy one of the old single Belfast hangars and on 22nd June 1968, this was done with incredible results, which achieved one of the highlights of the film. The film directed by Guy Hamilton and starring a multitude of famous actors including Christopher Plumber, Michael Caine, Sir Laurence Olivier, Trevor Howard and Suzanna York was a great success when it was released in 1969 and received many plaudits for its depiction of aerial combat.

In 1968, the Ministry of Defence put the site up for tender and during the next few years various plans for the redevelopment of the aerodrome were made. Cambridgeshire and Isle of Ely County Councils put forward proposals for a regional sports and recreation centre and country park, and the Home Office declared an interest in having two prisons with staff buildings.

By February 1971, a public enquiry into the proposed use of the aerodrome was undertaken with further proposals given by the East Anglian Aviation Society and The Imperial War Museum who sought to use the site as an aircraft museum within a recreation centre. Towards the end of that year an agreement was reached between the Ministry of Defence and other government departments that the proposed temporary use by the IWM and East Anglian Aviation Society, to use part of the aerodrome for historic aircraft to be housed, could go ahead. The Cambridge University Gliding Club was also given permission to use part of the aerodrome. Another aviation group also sought to help – the Shuttleworth Collection. The first aircraft exhibit to be flown in was a Royal Navy Sea Vixen which arrived in March 1972.

As other aircraft and historical exhibits arrived and the collection began to increase, the aerodrome was opened to the public for a special air day on 14th October 1973, which proved a great success. In 1975, the results of the public enquiry into Duxford were issued. The upshot was to give the Imperial War Museum full use of the aerodrome as an out-station for its collection. Cambridge County Council were licensed with running the site and the Department of the Environment were given the task of restoring buildings and hangars. Buildings of historical importance were to be given 'Listed' status. Sadly during this year, the East Anglian Aviation Society decided to move its collection out, due to irreconcilable differences with the Imperial War Museum. Those members who remained formed the Duxford Aviation Society, however.

On 29th June 1975, a two-day display was jointly organised by the Duxford Aviation Society and the IWM. The event was very well attended and included displays by five Pitt S2s of the Rothmans Aerobatic Team, a Fairey Swordfish, Sea Fury and a French Dewoitine D26.

By 1976, the IWM collection had grown considerably and it was decided in June 1976 to open to the public on a daily basis. Visitor numbers were better than expected and continued to rise. One drawback that year however was the acquisition of land by

Cambridge County Council for the building of the new M11 Motorway between London and Cambridge.

The land needed for the building of the motorway would take 150 acres from the aerodrome and shorten the usable runway by 1,500 feet. Finally, an agreement was struck that would see both interested parties happy with an outcome that was satisfactory.

One exhibit which arrived on 20th August 1977 and brought in many visitors and continues to do so now, was the first pre-production Concorde 01 G-AXDN, which touched down at Duxford amid much publicity and ceremony.

As more historic aircraft began to arrive for either display or restoration, the need for more storage and exhibit space became a priority. A new hangar was erected and this helped to ease the problem for the present. However, visitor numbers had increased to such an extent that by 1980, the figures had reached over a million.

Expansion of the aerodrome was put into plan, when it was decided to raise funds to build a 'super hangar' to house the larger aircraft like the Lancaster, Short Sunderland etc. In 1983, with substantial funds now available, the building contractors moved in to lay the foundations of the new hangar site. In 1985 the project was completed and was opened by His Royal Highness Prince Andrew, Duke of York.

During 1983, owners of private flying collections began to arrive to set up base at Duxford after agreement with the IWM. The collections included those of Stephen Gray known as the 'Fighter Collection', and Ray Hanna who founded the 'Old Flying Machine Company.' They have since become household names in aviation circles for their historic flying aircraft and restoration. Many of their aircraft have been used in recent movies including 'A Bridge too Far', 'A Piece of Cake' and 'Deep Blue World'.

Duxford was involved with another film project in 1989, when it was used for the movie 'Memphis Belle'; the story of an American bomber crew flying on its final mission. During the filming the aerodrome became home for five B17 Flying Fortresses, seven Mustang fighters and three Bf109s, which flew in and out during the day to other locations.

In 1990 plans to develop the western end part of the aerodrome got underway in order to house and display the military vehicles and tanks from the IWM and other contributors. The building would be titled the Land Warfare Hall. Organisations involved during the construction included British Steel PLC, the A.F. Budge Military Collection and Mabey & Johnson Limited. The building was completed in early 1992 and opened by Field Marshal, The Lord Bramall KG on behalf of the Prime Minister on 28th September 1992.

Plans for an American Air Museum to commemorate the achievements and the sacrifice of the United States Air Force during World War Two, had been put into effect during the early 1990s and architect Sir Norman Foster's design was selected. The contactors chosen for the project was John Sisk & Sons Ltd. Construction on the site started on 8th September 1995, with the final cost running to £11 million. The funds raised were from grants given from The Heritage Lottery, and business companies, as well as veteran groups and the general public.

On completion of the building, which is the largest unsupported concrete span in Europe, Her Majesty Queen Elizabeth officially opened the new American hangar on 1st August 1997.

Just inside the main entrance a plaque reads 'Dedicated to the memory of 30,000 American airmen who lost their lives from British bases in WW2.'

A rededication of the building was to take place, after further major improvements had been undertaken. The Honourable George Bush, 41st President of the United States arrived with his Royal Highness the Prince of Wales on 27th September 2002. Both dignitaries spoke of the special relationship and sacrifice that both countries had witnessed during WW2.

The latest project which is now nearing completion is the construction of the new Airspace building which will house and tell the story of British aviation from its earliest times. The building will have 12,000 square metres of space to display many of Britain's greatest aircraft like Concorde, the Avro Vulcan, Shorts Sunderland and many others. The building is being funded by the Heritage Lottery Fund who donated £9 million pounds and £995,000 from the East of England Development Agency. Other companies involved include BAE Systems in partnership with the Imperial War Museum. The new building is to be opened in 2007.

Duxford today maintains the role of Europe's top premier air show and aviation museum, with thousands of people continually visiting the aerodrome and this will continue to be so, for as long as the public's fascination with man's attempts to conquer the skies through war and peace remains.

THE BATTLE OF BRITAIN

Spitfire Productions Limited
4 South Street, London. W.1
Telephone: HYDe Park 7428/9

PLEASE REPLY TO:
PINEWOOD STUDIOS,
IVER HEATH, BUCKS.
TEL: IVER 700

21st February 1968

Mr. C. R. Smith,
Manor Farm,
Thriplow,
Cambridgeshire.

Dear Mr. Smith:

Following our site meeting at R.A.F. Duxford on the 8th of this month, I can now confirm the requirements of Mr. Guy Hamilton our Director, as follows:-
(For mutual reference I enclose a copy of a Plan of the location.)

(A) You have kindly consented the area of the Nissen Huts backed by the farm buildings may be used for filming and to represent a section of a French Farm and Airfield of the 1940 period.

(B) The area marked yellow and green may be used as a grass runway.

(C) The section marked yellow is your area of sewn barley which you have advised should be taken up and re-sewn with grass in order to achieve the effect of a grass runway.

(D) The track (marked in red) indicated near the end of the sewn barley area will be conditioned so that it would not impare the movement of aircraft in take-off or landing.

(E) The area East of the track (marked in Green) will be made into a grass runway as indicated.

(F) We have requested your permission to blend the area banded in blue by removing the part of the fence marked brown and by cutting the grass of both sides, also to re-decorate the front of the Nissen Huts and to place 'dressing' in this area that would represent the activity and temporary encampment of a Unit of the combined French Air Force and the Royal Air Force.

With reference to the newly proposed grass runway area, I confirm that we have requested you undertake this task on our behalf, which I summarise as follows:-

Cont'd/...

Directors: Harry Saltzman S. Benjamin Fisz K. J. Richards R. S. Aitkin

---

-2-

Mr. C. R. Smith: (cont'd)   21.2.1968

That the barley will be lifted. The whole area would be 'combed' (with special attention to the track mentioned above) be sewn by grass seed of your selection. The area would then borne by this Company and you will, at your earliest convenience, provide us with an estimate for the complete undertaking. The cost of this work will be discussion.

I trust that the foregoing coincides with your understanding of our discussion.

I should have pleasure in re-visiting you for further discussions whenever it is convenient to you or to discuss the matter further by telephone.

With sincere thanks for your co-operation,

Yours truly,

[signature]
PRODUCTION SUPERVISOR

Top left: The letter written to C.R. Smith by Spitfire Productions Ltd.

Above: Spitfire TR9 MJ772 (G-AVAV) comes in to land after another flight, during the making of the film. The aircraft was a two-seat variant. *(Courtesy of S. Saunders ASA Productions Ltd)*

Left: A Spanish Air Force CASA 2111 (Merlin-engine Heinkel 111) taxies along the runway at Duxford during filming. *(Courtesy of S. Saunders ASA Productions Ltd.)*

Left: Spanish Hispano Buchon Messerschmitt 109s lined up at the aerodrome prior to filming.
(Courtesy of Peter Arnold)

Below: The production unit for United Artists built this mock-up of a French chateau for their Battle of France airfield. This was filmed at the western end of the aerodrome.
(Jean Michel Goyat Collection)

Below: Aerial view of the aerodrome with Spanish Air Force Buchon Messerschmitts and a Heinkel bomber on left. Hurricanes and Spitfire are positioned at the bottom of the photograph. *(Courtesy of Peter Arnold)*

Above: A Messerschmitt dives down to strafe a Hawker Hurricane at dispersal in the early opening sequence for the Battle of Britain movie.
(Peter Arnold Collection)

Left: Hawker Hurricanes at the dispersal area ready for the French airfield sequence.
(Jean Michel Goyat Collection)

Left: One of Duxford's hangars was actually destroyed during the filming of the Battle of Britain. The destruction gave incredible realism to the film and recalled the German raid on the aerodrome. *(Peter Arnold Collection)*

Right: The remains of the hangar that was destroyed with replica Spitfire lying in pieces.

Left: Black smoke and earth rises skyward as pyrotechnics explode to recreate a Spitfire being blown apart during a Luftwaffe bombing raid. *(Courtesy S. Saunders ASA Productions Ltd)*

Above: The aerodrome on 28th/29th June 1975, when Duxford opened its gates to the public over two days. The event was well attended with the central exhibition area situated between the two hangars. Top right is the coach park. On the concrete apron was the live aircraft park with emergency vehicles in attendance.

Left: A party of school children are among the first visitors. *(Mh20239)*

Left: The Re-Dedication on Sunday 23rd May 1976, of the stone commemorating the service at Duxford of 78th Fighter Group, United States Air Force during WW2.

The stone had been in the safe keeping of the United States Air Force at RAF Lakenheath since 1970. In the photograph Colonel R.E. Messerli of 48th Tactical Fighter Wing unveils it along with J.W. Nelson a former 78th Fighter Group pilot. Also present from left to right, are Brian Biddulph-Pinchard, secretary of the Duxford Aviation Museum, Dr C.H. Roads, deputy director of the Imperial War Museum and M.J. Barnett, public relations officer of the Duxford Aviation Society. *(IWM HU 30997)*

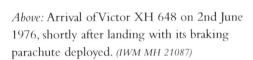

Above: Arrival of Victor XH 648 on 2nd June 1976, shortly after landing with its braking parachute deployed. *(IWM MH 21087)*

Right: Group Captain David Parry-Evans, station commander RAF Marham, and second pilot of the Victor with Dr C.H. Roads, deputy director of the Imperial War Museum who is looking at the pilot notes for Victor XH 648. On the left is Wing Commander Alistair Sutherland, officer commanding 57 Squadron, RAF Marham. *(IWM MH21093)*

Top left: Group Captain Sir Douglas Bader returns to Duxford to inaugurate the aerodrome's Vintage Air Day on 20th June 1976. *(IWM MH 21182)*

Top right: It's getting rather hot in here. A greenhouse is assembled on top of the roof of the control tower for use as the air traffic and display control centre for the Vintage Air Day. *(IWM MH 21175)*

Middle: The arrival of Concorde 01 on 20th August 1977. *(IWM MH 22884)*

Bottom: Dr C.H. Roads, deputy director, greeting the crew of Concorde 01, in particular Brian Trubshaw, director of flight testing, BAC. *(IWM MH 22890)*

Top: Three Spitfire replicas lined up alongside the hangars at Duxford during the Spitfire 50th Jubilee Air Show on 10th July 1986. *(Robert J. Rudhall Collection)*

Bottom: The inside roof structure and supporting columns of Duxford's Belfast Hangar 3, with the concertina-type doors to the right. *(Author)*

Top: Duxford's operations room now restored for the public to view as it would have been seen during the Battle of Britain. *(Author)*

Bottom: Field Marshal Bernard Montgomery's staff car, a Humber Mk 2 Super Snipe on display within the Land Warfare Hall. *(Author)*

The American Air Museum outside and in. Designed by Sir Norman Foster & Associates, it was opened on 1st August 1992.
(Author)

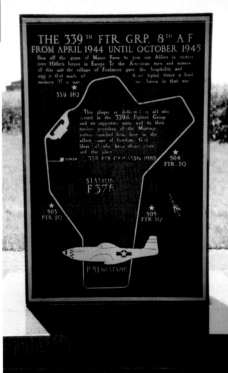

Top: Fowlmere today. A single American T2 hangar stands as a reminder of the airfield's wartime past. It is now a privately run airfield owned by aviation company Modern Air. *(Author)*

Bottom left: A mural adorns an end wall of the brick barn at Fowlmere. The RAF wings were painted in December 1940 by aircraftsman Robert Hofton for a Christmas dance.

Hofton was an artist in civilian life and was contacted by Martin Sheldrick in 1988, to restore the mural back to its original condition. *(Courtesy of Martin Sheldrick)*

Bottom right: The memorial dedicated to the 339th Fighter Group who served at Fowlmere from April 1944 until October 1945. *(Author)*

Top: Heroes return! Battle of Britain veterans are photographed together at the opening of the Battle of Britain Hall at Duxford to commemorate the 60th Anniversary of the battle in 2000.

Left to right back row: F/Lt L. Harvey, F/Lt R. Jones,?, F/Lt L. Martel, F/Lt O. Burns, W/Cdr J. Young, ?,?, W/Cdr T. Neil, Vivian Snell, S/Ldr R. Foster, S/Ldr T. Vigors. Front row left to right: ACM C. Foxley-Norris, W/Cdr J. Sanders, W/Cdr Wilf Sizer, F/Lt A.C. Leigh, W/Cdr G. Unwin, F/O K. Wilkinson, F/Lt T. Pickering, F/Lt A. Gregory, Air Marshal John Grandy, S/Ldr John Bentley-Beard, W/Cdr E. Barwell. *(Author)*

Bottom: Inside the Battle of Britain Hall, a Messerschmitt Bf109E which was shot down during the battle is displayed to the public in a depiction taken from an original photograph showing the crash-landing in 1940. *(Author)*

Top: Two distinguished veterans of the Battle of Britain catch up on old times in 2003. On the left is Squadron Leader Gerald 'Stapme' Stapleton with Squadron Leader Ben Bennions.

(Courtesy of Vector Fine Art)

Bottom: The Vector Fine Art stand at Duxford has raised thousands of pounds for Royal Air Force charities over the years. Pictured during one of Duxford's annual air shows are left to right: Battle of Britain veteran John Beazley, Colin Smith (proprietor) and Group Captain Patrick Tootal, chairman of the Battle of Britain Monument at Capel Le Ferne.

(Courtesy of Vector Fine Art)

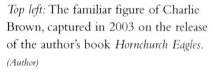

Top left: The familiar figure of Charlie Brown, captured in 2003 on the release of the author's book *Hornchurch Eagles*. *(Author)*

Top right: Under restoration. Gloster Gladiator N5903 was built in 1939, but was never used operationally during the war. In the 1970s it was loaned to the Fleet Air Arm Museum at Yeovilton and volunteers restored the aircraft back to static condition. In 1994, it was sold to The Fighter Collection who are currently rebuilding the aircraft to flying condition. *(Author)*

Middle: Members of the public gather during Duxford's Flying Legends Air Show on 9th July 2004 at the Vector Fine Art marquee for a Grub Street book launch and signing by WW2 veterans. *(Courtesy of Vector Fine Art)*

Bottom: The American Military Cemetery at Madingley, Cambridge, where 3,900 American servicemen lay at peace. *(Author)*

Looking to the future? The new Airspace Super Hanger due to open to the public in 2007. Positioned outside the hangar are Avro Vulcan XJ824 and BAC TSR2 XR222, the fourth TSR2 that was produced. *(Author)*

R.A.F. DUXFORD

1950 – CLOSURE

DID YOU SERVE THERE?

**OLD DUX ASSOCIATION WELCOMES ANYONE INTERESTED.
ALL RANKS AND TRADES**

MEETINGS BI-ANNUALLY

For details please phone Dartford (01322) 274245, or alternatively write to

Mr. J. F. Garlinge
45 Trevithick Drive
Dartford, Kent DA1 5JH

email Jfgalinge@aol.com